TABLE OF CONTENTS

ABOUT THIS BOOK

So you are looking at soy products! Welcome. It's an exciting food.

As technology for producing edible soy protein becomes more sophisticated, more and more people get interested in the product. It tastes better, it is easier to prepare and serve.

We still have a way to go.

This book is to be used as a source of ideas; suggestions of food combinations and flavors which enhance each other. Soy protein products may never escape their beany origins, but good "cover up" flavors and careful preparation will overcome the "beany" flavor.

Not so easy is the matter of deciding when to use which soy product—the higher cost protein isolate or the lower priced soy flour.

Recipes were tested with products as shown in ingredient listings. In most recipes soy protein concentrate can be substituted for soy protein isolate. Flavors and yields will be similar. However—the protein and other nutritional values will change; so will price.

Soy grits or soy flour are different-size particles made from the same soy product. Dry soy grits may be used successfully in most recipes calling for dry soy protein.

The most important factor to remember is that for each particular foodservice operation the requirements of customers will dictate choice of soy product.

Soy processors are plagued by flavors. The meat flavors used (because of lower cost) are easily dissolved in water. Long cooking times are not a good idea unless the liquid has a flavoring similar to the soy product. This means you use bouillon, au jus, beef stock, or gravy base in entrees. It means extra spices and seasonings.

Hope is around the corner. Oil-base meat flavors and encapsulated flavors are available from ingredient suppliers. The cost is still high and techniques for applying these flavors (making the flavor stick to the soy product) may have to be changed.

Be patient. Experiment. As my old professor kept telling us—"There are many ways to make a biscuit." There are no right or wrong ways to use soy protein—some are unusual and others may be easier.

Enjoy, enjoy.

THE PROFESSIONAL CHEF'S
SOY PROTEIN RECIPE IDEAS

NANCY SNIDER
Food Editor
Institutions/Volume Feeding Management Magazine

Jule Wilkinson
Book Editor

Tony Pronoitis
Art Director

Wilma Inman
Book Design

Published by
INSTITUTIONS/VOLUME FEEDING MANAGEMENT MAGAZINE
1801 Prairie Avenue
Chicago, Illinois 60616

Distributed to the book trade by:
Cahners Books, 221 Columbus Avenue, Boston, Mass. 02116

ISBN 0-8436-0540-5

Printed in the United States of America

SOY PROTEIN: FOOD SERVICE'S NEW STAPLE?

THE PROFESSIONAL CHEF

Your future begins right now! Soy protein, a major member of that class of man-made "new foods" is no longer a laboratory curiosity, plaything of the food technologist or lost in the backwater of the health food cult. It's now on the institutional food market in a variety of forms, flavors and textures, ready for incorporation into most menus.

The timing is right, too. The food service industry is facing a new set of challenges that demand a new look at traditional foods and traditional menus. There's the spectre of professional consumerism just around the corner . . . There's the vast commitment to feeding the nation's hungry well at low cost . . . There's the likelihood of new customer demands for greater nutritional values . . . and there is, suddenly, the prospect of soaring meat prices as a result of the Midwestern corn blight.

Because soy protein is such an important, new and different kind of ingredient, Institutions food editor Nancy Snider has spent the past two months working with the product in a test kitchen developing exclusive recipes . . . exploring soy's possibilities . . . overcoming its weaknesses. The special report on the pages that follow is the result. Comments of her test panel are included with each recipe.

Meats and poultry back up the newer soy proteins which can stretch diets and dollars. Shown on the butcher's block are, L to R: bacon flavored soy bits, ham flavored soy loaf, soya-meat fried chicken style, vegetable Skallops, all vegetable hot dogs, "sausage" soy, soy chunks with ham flavor, diced soy with beef flavor and gravy, soy chunks with chicken flavor.

A Harvest of 'New' Foods

Pea Soup •

Baked Ham   Harvest Hash •   Beef Stew •

Frankly Beany Bake •   Scalloped Potatoes •

Acorn Squash with Scrambled Soy "Sausage" •

Yams and Pineapple •   Corn on the Cob

Watermelon Pickles   Corn Relish   Applesauce

Sliced Tomatoes   Pickled Beets

Carrots   Radishes   Green Onions

Bread and Butter   Apple Jelly

Mince Pie •

Coffee   Iced Tea   Milk

•using soy protein products

"Dig in," said 'Aunt' Edith to the men on the Johnson 1910 Working Farm located outside Geneseo, Ill. And Institutions' soy recipes passed their final test: even in the heart of the farm country, diners went back for seconds.

The Soy Protein Fact Books

One of Man's Earliest Friends

Soybeans are one of the oldest crops cultivated by man. Chinese records dating back to 2000 B.C. mention use of cultivated soybeans; some historians say that China's survival has been possible because they used soybeans for food.

Soybeans were introduced into the United States in 1804. By 1924, 5 million bushels of soybeans were produced in the U.S. Today over 1 billion bushels are produced yearly in this country—more than 75% of the world's production.

At harvest time in late summer and early fall, soybeans contain about 38% protein and 18% oil. Protein is processed for edible products such as meat extenders, meat analogs, bakery ingredients. In the extended meat products, also called comminuted meats, government regulations on federal, state and local levels specify how much vegetable protein can be used. New York, for instance, allows "economy frankfurters" with 5% soy protein.

The Five Kinds of Soy Products

There are five cateories of modified soy products:

- Full-fat soy flour (up to 40% protein)
- Defatted soy flour (about 50% protein)
- A 60 to 70% protein concentrate
- Soy milk
- Isolated soy protein (90% protein)

Full fat soy products are made by heat-treating soybeans without removal of oil. Such products cost about 8 to 10 cents per pound, and go into beverages and baby foods.

Defatted soy costs least—about 9 cents per lb. This product is used in bakery and meat products, beverages and baby foods. Some textured products, such as ingredients for ground meat products or stews, are manufactured from this source.

Soy protein concentrates are made by removing carbohydrates, oil and the seed coat. These products cost more, about 25 cents per pound. They are useful for fortifying bread, for beverages and as the base material for textured proteins.

Soy milk is an extract of cooked soybeans. This product is usually sold only in health food stores.

Soy protein isolate is the modern version of the Oriental soy curd (tofu). It costs about 40 cents per pound. Special "functional properties" make it suitable for use in many food products because it can be spun into edible fibers, similar to the spinning of textile fibers.

The dry soy protein products used for these recipes cost 35-90¢ per lb. However, the product rehydrates to 3 times its original weight. The frozen product cost 83-87¢ per lb. canned product cost 61-90¢ a lb. Cost per pound of edible finished product becomes lower for soy protein than for many other proteins.

A Product for a Nutrition Conscious World

Soy protein products contain little roughage. An analysis for each product is usually available and generally shows low fat, low carbohydrate and high protein content. Moisture will vary greatly, depending on the form of product considered.

Methionine, an essential amino acid, is very low (almost at the "trace" level) in soy protein products. *Institution's* recipes use dairy products to increase the protein quality as well as quantity.

The carbohydrates which causes flatulence (raffinose and stachyose) are removed from soy protein concentrates but are present in the soy flours and grits used as ingredients in a variety of products.

Little or no cholesterol is present in most soy protein products. Some processors add animal fats for flavoring, however.

Soy by Any Other Name

What do you call soy protein on your menu? It's an important consideration with consumerism rampant in the land.

How about an "all vegetable protein plate" in which you use "beans" or "protein" to describe the soy protein product.

If you are using the product of one manufacturer, perhaps his brandname, or trademark, can be used on your menu: "creamed Bontrae with a flavor l i k e chicken," "soyameat fried chicken style."

F o r Institutions Magazine's recipes, people's names and vegetable ingredients were used rather than the meat names. By eliminating meat names, there is less likelihood of consumer complaints on mislabeling. Each dish needs one or two lines of explanation of what soy protein is and where it came from.

Meanwhile, Back in The Laboratory...

Food technologists across the country are extruding, puffing and toasting soy proteins to m a k e snack products; flavors range from garlic to strawberry. A cautious kind of replacement is continuing; soy protein in the form of grits and flour replace some of the wheat flour and eggs in baked products. For a penny more on ingredient costs, the 12-oz. bottle of cola could be equal to milk in protein.

Our traditional approach to foods is the "meat and potatoes" route. As operators get to the point of seeing new foods on the basis of what they do for people, what function in nutrition they fulfill, then we are ready for our foods of the future right now.

Food chemists say, "We shall overcome —the beany flavor." Food-service operators must say, "We shall overcome the natural resistance to change—by making soy protein and other new foods so attractive that customers flock in the door to try the marvelous new food.'

Soybeans, meat analogs, will never replace real meat. But—the time is almost upon us when meat will not be as plentiful as we would like. Soy protein products are a useful supplement, not a cure-all for the population explosion or the cost-price squeeze.

Canned, frozen or dry soy protein offers storage convenience. Shown, L to R: dry soy flakes, frozen "sausage" soy, canned soy hot dogs, diced soy with beef flavor and gravy, soy beans, frozen soy chunks with ham flavor and frozen soy chunks with chicken flavor.

Soy What?

Soy protein is similar to meat in that:
it retains its integrity—if it starts out as a chunk, very little that you can do to it in cooking will change its basic shape and texture

It is different from meat in that:
it does not shrink
it has little or no fat (usually)
it has a faint nutty flavor of its own, and only with difficulty and care have companies been able to make other flavors "stick" to it
the flavor with which soy protein comes into the kitchen can be easily washed off, soaked off and sometimes baked off

Canned Soy
although the liquid in the can has helped give the soy protein its flavor, usually it is best discarded and replaced by a sauce or gravy
the "beany" flavor sometimes associated with soy is more pronounced if the soy product liquid is used in cooking, and for the same reason, these recipes often recommend rinsing after draining the soy product

Frozen Soy
these products handle well either frozen or thawed, but care in reading the recipe must be used to discover if the weighing and cooking can be done while it is frozen or thawed they may become rancid because animal fats are used along with the soy protein

Dry Soy
a joy to have in the kitchen if well flavored
is light weight and will rehydrate to up to 3 times the original weight
wine, beer, vinegar, lemon juice or tomato juice should be used sparingly to rehydrate, but can be used in cooking— water, consomme or bouillon do best for reconstitution or rehydrating
little problem with storage if kept in tightly covered container in cool dry place

Try Ethnic Soy Dishes

Soy protein can be prepared with any kind of equipment—even an old cookstove. Here Aunt Edith tastes a gumbo.

Bombay Korma

YIELD: About 1 gal.

INGREDIENTS:

Tomato puree	29 oz. can
Water	1 qt.
Powdered milk	1 cup
Dehydrated onion flakes	½ cup
Margarine	¾ lb.
Cumin	1 tsp.
Ginger	1 tsp.
Turmeric	1 tsp.
Cardamon	¾ tsp.
Cinnamon	¾ tsp.
Coriander	½ tsp.
Powdered garlic	½ tsp.
Paprika	½ tsp.
Salt	½ tsp.
Tabasco	½ tsp.
Cloves	¼ tsp.
White pepper	¼ tsp.
Frozen chicken soy protein chunks	3 lb.

METHOD: Combine tomato puree, water, powdered milk and onion flakes. Set aside. Melt margarine in saucepot. Add spices and seasonings. Stir in tomato mixture. Simmer 30 min. Add soy protein. Simmer 30 min. more. Serve over rice.

COST BONUS: Substitute 2 (45 oz.) cans soyameat fried chicken style or 1 (4 lb. 12 oz.) can vegetable Skallops for frozen soy protein.

FOR GREATER FLAIR: For either canned soy protein product, bread lightly and deep fry. Place in 20 x 12 x 4 in. pan. Pour sauce over. Bake at 375F for 20 min.

Bombay Korma
"You mean ALL those spices go in there?"—secretary

"Well, my roommate likes curry and so do I. The dry milk in there helps mellow the flavors but it IS spicy."—an editor

Ah-So Chow Mein

YIELD: 9 qt.

INGREDIENTS:

Margarine	4 oz.
Celery	1 lb.
Dehydrated onion flakes	3 oz.
Dry hamburger soy protein granules	24 oz.
Chop suey vegetables 4 (3 lb. 2 oz.) can	
Water	1 qt.

METHOD: Melt margarine in large skillet. Add celery and onion flakes. Saute until celery is bright green. Then combine all ingredients in steam jacketed kettle. Cook 30 min.

Ah-So Chow Mein
"You really don't need au jus base in there for me."—engineer

Bombay Korma is a mystical curry made with tomato sauce to encourage orders of large bottles of sparkling rose or cold ale. Table-side presentation would include "boys" and saffron rice.

Soy with a Spice

Hot Pepper Quiche

YIELD: 3 pans (12x10 in.) or 5 pies (9 in.) or 2 pans (11x17 in.) or 21 individuals (4½ in.)

INGREDIENTS:

Pastry dough	About 3 lb.
Chopped onions	9 oz.
Shredded hot pepper cheese	9 oz.
Frozen ham soy protein chunks, thawed	1 lb.
Milk	2 qt.
Eggs	20 oz.
Salt	1½ tsp.
Chopped chives	

METHOD: Roll out pastry dough to fit pans. Line pans and flute edges. Portion out onions and cheese; spread over pastry in pans. Beat eggs, milk and salt just until blended. Portion into each pan. Portion soy protein into each pan. Sprinkle chives over top. Bake at 375F on bottom rack of oven; 12 x 10 in. or 11 x 17 in. pans for 40 min. to 1 hour; 9 in. pies for 35-45 min.; individual pies for 15-25 min. or until knife inserted 2 in. from edge comes out clean.

COST BONUS: Substitute reconstituted dry ham soy protein granules for frozen soy and add 1 Tbsp. sugar to egg mixture.

Hot Pepper Quiche
This was a great hit at a party—the men liked it because it was tangy; the women thought it shouldn't be fattening (they were wrong, but it was a nice illusion)

Scalloped Potatoes

YIELD: 20x12 in. pan, 24 servings

INGREDIENTS:

Dehydrated shredded potatoes	2½ lb.
Canned diced celery	1 pt.
Canned diced onions	1 pt.
Dehydrated diced green pepper	1½ oz.
Salt	2 Tbsp.
Pepper	1 tsp.
Shredded cheese	20 oz.
Frozen ham soy protein chunks	24 oz.
Milk	3 qt.
Margarine, melted	4 oz.

METHOD: Rehydrate potatoes as directed on package. Combine all ingredients in steam jacketed kettle; cook 15 min. Turn into 20 x 12 x 6 in. pan. Pour melted margarine over top. Bake at 400F for 20-30 min.

COST BONUS: Substitute reconstituted dry ham soy protein chunks for frozen soy protein. Stir often during baking.

Scalloped Potatoes
An old stand-by—where the variety is in what kind of cheese you use—this recipe used the hot pepper cheese, but a cheddar or process American cheese are OK too.

By making this recipe again with "real" ham, and mixing the 2 in the baking pans, you have an example of the 50:50 ratio.

Harvest Hash

YIELD: 24 servings

INGREDIENTS:

Diced beef soy protein	4 lb. 12 oz. can
Frozen french fries	2½ lb.
Canned corn	1 qt.
Brown gravy	3 pt.
Worcestershire sauce	1 Tbsp.
Salt	1 tsp.
Pepper	½ tsp.
Pre-prepared 101 Sauce or All-Purpose Tomato Sauce	1 cup

METHOD: In 20 x 12 x 4 in. pan, combine soy protein, french fries and corn. Mix gravy with 101 Sauce and remaining ingredients. Pour over top. Bake at 400F for 30 min.

COST BONUS: Rehydrate 20 oz. dehydrated shredded potatoes; substitute for the french fries. Rehydrate 1½ lb. dry beef soy protein chunks or hamburger granules in 1 pt. cola soft drink and 46 oz. can tomato juice. Substitute 2 pts. water for equal amount of gravy. Stir several times during baking.

Harvest Hash
Hash is a "basic standby" in which leftover vegetables and less expensive potatoes (such as the irregular cuts of french fries) can be used.

"I like it with biscuits and not so much gravy."—art director (male)

"I like more gravy and no topping so I don't have to worry about calories."—editor (female)

Hot Pepper Quiche offers profitable variety in size and shape for appetizer or petite dejeuner (small, light meal such as luncheon or midnight supper.)

Cooking with Soy Protein

Beefing Up Recipes with Soy Products

Recipes can be "extended" or increased in quantity without as great an increase in cost, by using soy protein.

It is best to use half or less soy protein. For meat/soy patties or loaves, 3% up to 30% could be soy. The great advantages are that products are juicy and shrink very little.

In ground meat mixtures such as taco filling, pizza topping or chili, half ground beef and half soy protein work very well. For extra flavor, substitute pork sausage for some ground beef. In some pre-prepared chili all soy is used, but the flavor is better with half meat. Try the rehydrated dry hamburger-flavored soy.

If your stew recipe calls for long slow cooking, use 5-50% dry soy protein. Rehydrate the soy and add during the last hour of cooking. This avoids a "cardboard" flavor which happens when the flavoring soaks off the soy into a bland sauce.

Unflavored soy can be added to tuna fish dishes. With 20-30% soy, the tuna flavor is mellow but not lost. An "ingredient" soy product is best for this use, such as minced soy protein concentrate.

Secrets of Soy Protein Cookery

Sauces, the spicier the better, are excellent with soy protein products. Some of the flavor from the sauce blends into the soy (which has only a slight flavor).

You can treat as a problem or a challenge—the fact that during cooking soy protein tends to lose the flavor which the processor adds. There seems no way to make soy products as rich in flavor as meat. Mild flavor, not delicate flavor, is the rule.

Au jus, bouillon and monosodium glutamate (MSG) are flavor enhancers which often can be used to good advantage with soy protein products.

All-soy products are tough or tender, depending on the texture desired by the manufacturer. There is little to do in the kitchen to change this characteristic. However, it is all too easy to dry out a soy protein product and achieve a tough, ugly little ball of misery which cannot be served because it simply cannot be eaten.

Saute, but don't brown, soy protein. Seared or charred soy protein tastes almost as bad as burned cabbage. Even baking can cause a problem unless the product is stirred frequently; stir soy protein casseroles every 10 or 15 minutes during baking.

Preparation and Equipment Techniques for Soy

A ham-flavored soy protein loaf is best used for sandwiches. Here's an idea—this product comes frozen, so slice very thin. Then make sandwiches and heat in the microwave oven. Pre-prepared cheese sauce, right from the can, can be spooned over the loosely piled sliced loaf on day-old bread. Heat in the microwave oven for 15-20 seconds. Try your own version with rarebit or a cream sauce.

The reason microwave cooking is good with soy protein loaf is that there is little or no fat in the soy product, and the moisture is brought to the surface, making the product taste very juicy and flavorful. This is also what happens in the deep fried soy protein products which are re-heated in a microwave oven.

Soy products do well in most conventional equipment. In the convection oven frequent stirring is an absolute for soy protein casseroles. The steamer is useful for pre-prepared products, such as a meat loaf, frozen and ready to reconstitute and serve. A tilt-fryer is recommended for sauteing vegetables; add the soy protein and sauce and simmer gently for an Oriental-type dish. The steam jacket kettle and regular saucepans will see the most use for your soy protein cookery. A deep fryer can help in unexpected ways. All-soy hot dogs, deep fried for a minute at 375F, taste better in baked beans.

All-Purpose Tomato Sauce

YIELD: About 5 pt.

INGREDIENTS:

Tomato puree	2 (29 oz.) can
Canned chopped celery	1 cup
Canned chopped onion	1 cup
Sweet pickle relish	1 cup
Red wine	1 cup
Dehydrated diced green pepper	¼ cup
Salad oil	¼ cup

METHOD: Combine all ingredients in saucepot. Cook, stirring often, about 45 min. Can be stored in jars in refrigerator for 10 days.

NOTE: Recipe may be increased 2 or 3 times without adjustment.

All-Purpose Tomato Sauce
This recipe is meant to take the place of a specific brand of sauce.

It's simple to make and store; you can vary it by adding spices, vinegar or sugar.

Drunken Soy
METHOD: Marinate drained and rinsed soyameat fried chicken style in white wine to cover for 1 to 2 hours. Serve hot with sauce or chill to serve with salad plates.

"Say, that's the best chicken I've tasted. A little moist though. Soy? You're joking. Well, I certainly think it has to say so on a menu if I'm eating out."—secretary

Tempura Style
METHOD: Prepare tempura batter using sherry instead of water. Coat marinated chicken soya, fry in deep fat at 375F until golden, about 3 min.

The Maharaja's Divan
YIELD: 24 servings
INGREDIENTS:

Frozen broccoli spears	2 (2 lb.) pkg.
Soyameat fried chicken style	2 (2 lb. 13 oz.) cans
Melted margarine	4 oz.
Salt	1 Tbsp.
Pepper	1 tsp.
Cheese sauce	1 qt.
Parmesan cheese	2 oz.

METHOD: Cook broccoli as directed on package. Arrange spears in 20 x 12 x 4 in. pan so flower ends are toward sides of pan and stalks are toward center. Combine margarine, salt and pepper; pour over broccoli. Drain and rinse soya chicken. Arrange down center of pan. Spoon dollops of cheese sauce over top. Sprinkle with Parmesan cheese. Bake at 400F just until sauce bubbles, about 10 min.

NOTE: Sherry Sauce may be substituted for cheese sauce.

Sherry Sauce
METHOD: Prepare 1 qt. classic medium cream sauce. Stir in ½ cup sherry. Bring just to a boil.

A combination of broccoli, sauce and poultry is traditionally considered a "divan." For the Maharaja's Divan, the broccoli is steamed just until tender-crisp.

After broccoli is arranged in the steam table pan, soya-meat fried chicken style goes on top. Appearance plays a big part in appetite appeal.

Spoon canned cheese sauce over soy and broccoli.

Parmesan cheese is sprinkled over the top for eye appeal as well as for enhanced flavor.

The big test is: will the customer buy it? Even a free lunch has to taste good and look pretty, especially if the tasters are also testers.

WHO MAKES WHAT

Company	Trade name	How used in recipes
Archer Daniels Midland Co.	TVP (textured vegetable protein)	dry soy protein
*Central Soya	Promosoy	
*Farm-Mar-Co.	Ultra-Soy	
General Mills, Inc.	Bac*os	dry soy protein
	Bontrae	frozen soy protein
*The Griffith Laboratories, Inc.		
*Ralston Purina Co.	Edi-Pro	
*A. E. Staley Co.	Mira-Tex	
*Swift & Co.	Texgran	
*H. B. Taylor Co.	Textrasoy	
Worthington Foods, Inc.	Fibrotein	dry soy protein canned soy protein

*These companies make several forms of soy protein ingredients, but do not market any "finished" products to use in recipes.

This is the spicy, slightly sweet, rich tomato sauce so often used in
Middle European cooking. Other types of soy protein could be used
in it, or a combination of meat and soy.

Give Soy an Italian Accent

Truly beautiful to eat, Tony's Spinach Lasagna combines the traditional spinach of northern Italy with a Roman tomato sauce.

Popadic's Spicy Tomato Sauce

YIELD: About 3 pt.

INGREDIENTS:

Pre-prepared 101 Sauce or All-Purpose Tomato Sauce	1 pt.
Water	1½ cup
Frozen ham or chicken soy protein chunks	1 lb.
Bottled lemon juice	2 Tbsp.
Sugar	1 Tbsp.
Bay Leaf	1
Cinnamon	½ tsp.
Salt	½ tsp.
Nutmeg	¼ tsp.
Tabasco	¼ tsp.
Cloves	⅛ tsp.

METHOD: Combine all ingredients in large saucepan. Simmer 30-40 min. Serve over cooked pasta or cabbage rolls.

NOTE: Recipe can be increased 2 or 3 times without adjustment.

Tony's Spinach Lasagna

YIELD: 12x20 in. pan, 18-24 servings

INGREDIENTS:

Eggs	12 oz.
Bottled lemon juice	2 Tbsp.
Salt	1 Tbsp.
Nutmeg	1 tsp.
Pepper	1 tsp.
Frozen chopped spinach, thawed and well drained	2 lb.
Dry bread crumbs	6 oz.
Parmesan cheese	6 oz.
Creamed small curd cottage cheese	2 lb.
Wide lasagna noodles	12 oz.
Margarine or shortening	4 oz.
Terry's Sauce	½ recipe

METHOD: Beat together eggs, lemon juice and spices. Combine spinach with bread crumbs, 2 oz. Parmesan and 4 oz. egg mixture. Combine cottage cheese with 2 oz. Parmesan and remaining egg mixture. Cook noodles in salted boiling water. Drain and rinse. Grease 12 x 20 x 4 in. pan with the margarine.

Make a layer of noodles on bottom of pan. Spread spinach mixture evenly over noodles. Place another layer of noodles on top. Cover with cottage cheese mixture. Place third layer of noodles on top. Cover with soy protein portion of Terry's Sauce. Place remaining noodles on top. Mix Parmesan with remaining sauce and spread over noodles. Bake at 400F for 30-40 min.

Terry's Sauce for Lasagna

YIELD: About 1 gal.

INGREDIENTS:

Pre-prepared 101 Sauce or All-Purpose Tomato Sauce	2 qt.
Water	1 qt.
Mushrooms	2 (8 oz.) can
Chopped parsley	2 Tbsp.
Oregano	4 tsp.
Sweet basil	2 tsp.
Frozen ham soy protein chunks	3 lb.

METHOD: Combine all ingredients except soy protein. Simmer 20 min. Set aside 3 pt. Add soy protein to remaining sauce; simmer 10 min. longer.

COST BONUS: Substitute 1 lb. dry ham soy protein granules for frozen soy. Increase water to 3 qt. and simmer soy protein in sauce over very low heat for 30 min.

Or, substitute 1 (4 lb. 12 oz.) can beef soy protein for the frozen soy. Drain and rinse before adding to sauce.

Soy for More than Entrees

Corn Muffins with bacon-flavored soy bits can be served for home-style breakfasts, with luncheon specials or in snowy napkins for dinner.

Rama's Curried Fruit

YIELD: About 6 pt.

INGREDIENTS:

Fruits for salad, well drained	3 (30 oz.) can
Fresh blueberries	1 pt.
Mayonnaise	1 pt.
Curry powder	1 Tbsp.
Bottled lemon juice	1 Tbsp.
Garlic powder	½ tsp.
Ground ginger	½ tsp.
Tabasco	¼ tsp.
Rum sour or daiquiri	8 oz.
Frozen chicken soy protein chunks, thawed	1½ lb.

METHOD: Combine fruits, mayonnaise, seasonings and Tabasco. Cover and chill. Marinate soy protein in rum cocktail at least 30 min., but no longer than 2 hours. Then combine soy protein with fruits. Serve well chilled, with "boys."

COST BONUS: Substitute 2 cans fruit cocktail for part of fruit. Substitute sliced fresh apples for blueberries.

Extended Tuna Salad for Monte Cristos

YIELD: 8 sandwiches

INGREDIENTS:

Tuna	9¼ oz. can
Minced unflavored soy protein	3 oz.
Chopped green pepper	½ cup
Mayonnaise	½ cup
Sweet pickle relish	⅓ cup
MSG	½ tsp.
Bread	16 slices
Egg wash	1½ cup
Corn flake crumbs	3 cup

METHOD: Combine tuna, soy protein, green pepper, mayonnaise and relish. Chill for at least 8 hrs. Spread about ¼ cup tuna mixture on each of 8 slices bread. Top each with second slice. Dip each sandwich in egg wash, then in crumbs. Deep fry at 375F until golden brown, about 3 min.

Corn Muffins

METHOD: For each lb. dry muffin mix, add 1½ oz. bacon soy protein bits.

Rama's Curried Fruit calls for "boys" of chutney, coconut, dilled onion rings, light and dark raisins and bacon-flavored soy bits. Peanuts and seasoned almonds would be good too.

Soy Helps Hold Down Costs

The extent of the damage caused by the blight (above) that swept across the midwest cornbelt can't be determined until the full crop is harvested. But soy proteins could be an important factor in helping the foodservice industry fight possible rising costs.

Few industries are required to cope with so many totally unexpected variations in the cost of their raw material as is foodservice.

Take our basic raw material—food.

Because of the corn blight which swept through the midwest this summer, the corn yield is expected to be significantly smaller than in the recent past. The upshot could be an increase in the price of meat, poultry and eggs that some experts say could go as high as 30%.

Fortunately for the foodservice industry, the development of new foods such as soy proteins may help offset this increase in the cost of our raw material.

Institutions' *feature on soy protein was started long before the blight began spreading across the cornbelt. But the testing we did all summer assumes special significance now that necessity may accelerate the use of the products.*

In considering and testing soy proteins for introduction into your own operation, we hope you can benefit from our experience:

1. The operator must be prepared to overcome his (and his kitchen employees') negative first reaction to working with the product. When it arrives in your kitchen, much of it bears a remarkable similarity in appearance to pet food.

2. In spite of this drawback, soy protein produces excellent results. The most constantly repeated comment during the months we tested was, "I had no idea it would be so good. This really tastes appetizing!"

3. Unless your food budget is very low, *we suggest using soy protein with real meat or poultry in not more than a 50-50 ratio. Our staff felt that up to this point it was difficult to tell the difference from the traditional item. If more soy was used, the item tasted "different." This is especially important in dishes that need gravy. Soy protein does not make gravy when cooked, and we found the liquid in which it is packed a poor substitute.*

4. As with most products, we found the higher priced soy products more acceptable than the cheaper varieties.

5. If you are hesitant about using the product in your own kitchen, check with your meat purveyor. Most major meat packers either have (or will soon have) products available that already combine soy protein with meat. This is particularly common in the case of meat patties, chili, etc.

6. We found patron acceptance increased when we mentioned the fact that soy beans are a natural (not synthetic) food to those who tasted our formulations.

In addition to the recipes given in this section (all developed exclusively for Institutions *by food editor Nancy Snider), there are many additional recipes available from manufacturers of soy protein products. To receive a list of names and addresses of companies with recipes available, just circle Number 750 on the Information Retrieval Card at the back of this issue.*

—Jane Wallace

ACKNOWLEDGEMENTS

Many people and companies have been most helpful: in supplying products to use in recipe development; others have contributed ideas and critiques (as well as having eaten their way through the recipes); Mrs. Harlow, who typed the manuscript did a painstaking job.

Jule Wilkinson, as editor, encouraged and corrected and generally pulled this book together.

The Food Protein Council provided the colorful cover, Doug Ross, photographer.

To all of you—thank you.

Also listed below are the companies whose product names appear on these pages as well as others whose helpful suggestions have been incorporated in this book.

Advisory Council for Jams, Jellies and Preserves

American Spice Trade Association

Angostura-Wuppermann Corp.

Archer-Daniels-Midland Co. (ADM)

Bordo Products

CAFE Program, George Westinghouse High School, Chicago

California Raisin Advisory Board

Central Soya Co., Inc., Chemurgy Div.

Cling Peach Advisory Board

Continental Coffee Co.

Durkee Food Service Group, Glidden Durkee

Durum-Macaroni, HRI Program

Food Protein Council

General Foods Corp.

General Mills, Inc.

Grocery Store Products

H. J. Heinz Co.

Idaho-Washington Dry Pea & Lentil Commissions

Institutions/VFM Magazine Editorial Staff

Land O' Lakes, Inc.

Libbey Products, Owens—Illinois

Libby, McNeil & Libby

Louisiana Yam Commission

National Broiler Council

National Cherry Growers & Industries Foundation

National Kraut Packers Association

North American Blueberry Council

The Olive Administration Committee

Pacific Coast Canned Pear Service

The Pillsbury Co.

Purple Plums Northwest

The Quaker Oats Co.

Ralston Purina Co.

ReaLemon Foods

Rice Council for Market Development

A. E. Staley Co.

S and W Fine Foods

Swift Chemical Co.

U. S. Brewers Assn.

Wheat Flour Institute

Worthington Foods, Inc.

SPICE INFORMATION CHART*

SPICE & SOURCES	DESCRIPTION	FLAVOR	SPICE EQUIVALENTS TEASPOONS PER OUNCE
ALLSPICE: Jamaica Honduras Mexico	Reddish-brown berries, nearly globular; 1/8 to 5/16-in. diam. Available: whole and ground.	Pungent, clove-like odor and taste.	14
ANISE SEED: Spain Netherlands Mexico	Greenish-brown, ovoid-shaped seeds, 3/16-in. long. Available: whole and ground.	Pleasant, licorice-like odor and taste.	14-1/2
BARBECUE SPICE: Mfd. U. S.	Blend of such as chili peppers, cumin, garlic, cloves, paprika, salt and sugar.	Characteristic, aromatic odor with varying levels of pungency and sweetness, according to brand.	12
BASIL: California Hungary France Yugoslavia	As marketed, small bits of green leaves. Available: whole and ground.	Aromatic, faintly anise-like; mildly pungent taste.	35
BAY LEAVES: Turkey Portugal	Elliptical leaves, up to 3-in. long; deep green upper surface, paler underneath. Available: whole and ground.	Fragrant, sweetly aromatic; slightly bitter taste.	136 leaves
CARAWAY SEED: Netherlands Poland	Curved, tapered brown seeds, up to 1/4-in. long. Available: whole.	Characteristic odor; warm, slightly sharp taste.	9-1/2
CARDAMOM SEED: Guatemala India	Small, angular, reddish-brown seeds; often marketed in their pods—greenish or buff-colored (blanched). Available: whole, decorticated and ground.	Pleasantly fragrant odor; warm, slightly sharp taste.	14-1/2
CELERY FLAKES: California	Medium to dark green flakes, about 3/8-in. diam. Available: flakes, granulated and powdered.	Sweet, strong typical celery odor and taste.	35
CELERY SEED: India France	Grayish-brown seed, up to 1/16-in. diam. Available: whole, ground and salt.	Warm, slightly bitter celery odor and taste.	14
CHILI POWDER: California	Red to very dark red powder.	Characteristic, aromatic odor with varying levels of heat or pungency.	11-1/2
FREEZE-DRIED CHIVES: California	Bright green, cross-cut sections of the tubular shoots, about 1/8-in. long.	Mild, delicate onion odor and taste.	160

*Chart courtesy of American Spice Trade Assn., New York City covers spices used in recipes in this book.

SPICE & SOURCES	DESCRIPTION	FLAVOR	SPICE EQUIVALENTS TEASPOONS PER OUNCE
CINNAMON: Indonesia Seychelles Ceylon Taiwan,	Tan to reddish-brown quills (sticks) of rolled bark, varying lengths. Available: whole and ground.	Agreeably aromatic with sweet, pungent taste.	17-1/2
CLOVES: Malagasy Republic Indonesia Tanzania	Reddish-brown, 1/2 to 3/4-in. long. Available: whole and ground.	Strong, pungent, sweet odor and taste.	14-1/2
CORIANDER SEED: Morocco Rumania Argentina	Yellowish-brown, nearly globular seed; 1/8 to 3/16-in. diam. Available: whole and ground.	Distinctively fragrant, lemon-like odor and taste.	14
CUMIN SEED (Cumino): Iran India Lebanon	Yellowish-brown, elongated oval seeds; 1/8 to 1/4-in. diam. Available: whole and ground.	Strong, aromatic, somewhat bitter.	14
CURRY POWDER: Mfd. U. S.	Blend of 16 to 20 spices, i.e. ginger, turmeric, fenugreek, cloves, pepper, cumin, cinnamon, etc.	Pleasantly fragrant odor, characteristic of Indian curry dishes, moderately warm to hot taste according to blend.	12-1/2
DILL SEED: India	Light brown, oval seeds; 3/32 to 3/16-in. long. Available: whole and ground.	Clean, aromatic odor; warm, caraway-like taste.	14
FENNEL SEED: India Argentina	Green to yellowish-brown seeds, oblong oval; 5/32 to 5/16-in. long. Available: whole and ground.	Warm, sweet, anise-like odor and taste.	14
GARLIC, DRIED: California	White material ranging in standard particle size from: powdered, granulated; ground; minced; chopped; sliced; large sliced.	Strong, characteristic odor, extremely pungent taste.	(Powder 15 (Salt 7 (Minced 10
GINGER: Nigeria Sierra Leone Jamaica	Irregularly shaped pieces ("hands") 2-1/2 to 4-in. long; brownish to buff colored (when peeled and bleached). Available: whole, ground, cracked.	Pungent, spicy-sweet odor; clean, hot taste.	14
ITALIAN SEASONING: Mfd. U. S.	Blend of such as oregano, basil, rosemary.	Highly aromatic, mildly pungent taste.	26
MACE: Indonesia Granada	Flat, brittle pieces of lacy material, yellow to brownish-orange in color. Available: whole and ground.	See nutmeg; but somewhat stronger, less delicate.	14

SPICE INFORMATION CHART (cont.)

SPICE & SOURCES	DESCRIPTION	FLAVOR	SPICE EQUIVALENTS TEASPOONS PER OUNCE
MARJORAM: France Portugal Greece Rumania	As marketed, small pieces of grayish-green leaves. Available: whole and ground.	Warm, aromatic pleasantly bitter, slightly camphoraceous.	19-1/2
MIXED VEGETABLE FLAKES: Mfd. U. S.	Typically composed of flakes of onion, celery, green and red peppers, and carrots.	See individual descriptions of these ingredients.	22
MUSTARD: Canada Denmark U. K. U. S.	Tiny, smooth, nearly globular seeds, yellowish or reddish-brown. Available: whole and ground.	Yellow; no odor, but sharp, pungent taste when water is added. Brown: with water added, sharp, irritating odor, pungent taste.	14-1/2
NUTMEG: Indonesia Granada	Large brown, ovular seed; up to 1-1/4-in. long. Available: whole and ground.	Characteristic, sweet, warm odor and taste.	12-3/4
ONION, DRIED: California	White material ranging in particle size from: powdered; granulated; ground; minced, chopped; large chopped; sliced; large sliced.	Sweetly pungent onion odor and taste.	(Powder 18 (Salt 8 (Minced 16 (Flakes 18
OREGANO: Greece Mexico Japan	As marketed, small pieces of green leaves. Available: whole and ground.	Strong, pleasant, somewhat camphoraceous odor and taste.	26
PAPRIKA: California Spain Bulgaria Morocco	Powder, ranging in color from bright, rich red to brick-red, depending on variety and handling.	Slightly sweet odor and taste; may have moderate bite.	13-1/2
PARSLEY FLAKES: California Texas	Flakes of bright green leaves. Also granulated.	Characteristic, mild agreeable odor and taste.	50
BLACK PEPPER: Indonesia Brazil India Malaysia	Brownish-black, wrinkled berries; up to 1/8-in. diam. Available: whole ground, cracked and decorticated.	Characteristic, penetrating odor; hot biting taste.	15-1/4
WHITE PEPPER: Indonesia Brazil Malaysia	Yellowish-gray seed, up to 3/32-in. diam. Available: whole and ground.	Like black pepper, but less pungent.	13-1/2
POPPY SEED: Rumania Turkey Netherlands Poland	Tiny round seeds, slate blue in color. Available: whole.	Mild, nut-like odor and taste.	11-1/2

*Chart courtesy of American Spice Trade Assn., New York City covers spices used in recipes in this book.

SPICE & SOURCES	DESCRIPTION	FLAVOR	SPICE EQUIVALENTS TEASPOONS PER OUNCE
RED PEPPER: Japan Mexico Turkey U. S.	Elongated and oblate-shaped red pods of varying sizes, from 3/8 to 12-in. depending on variety. Available whole and ground.	Characteristic odor with heat levels mild to intensely pungent.	(Ground 14 (Crushed 16
POULTRY SEASONING: Mfd. U. S.	Blend of such as sage, thyme, marjoram, savory; sometimes rosemary and other spices.	Highly aromatic with warm, slightly bitter taste.	26
ROSEMARY: France Spain Portugal California	Bits of pine needle-like green leaves. Available: whole and ground.	Agreeable, aromatic odor; fresh bittersweet taste.	35
SAFFRON: Spain Portugal	Orange and yellow strands, approximately 1/2 to 3/4-in. long. Available: whole and ground.	Strong, somewhat medicinal odor; bitter taste.	35
SAGE: Yugoslavia Albania	Oblate-lanceolate shaped leaves, grayish-green, about 3-in. long. Available: whole cut, rubbed, ground.	Highly aromatic, with strong, warm, slightly bitter taste.	22
SAVORY: France Spain	As marketed, bits of dried greenish-brown leaves. Available: whole and ground.	Fragrant, aromatic odor.	18-1/2
SESAME SEED: Mexico Nicaragua Guatemala Salvador U. S.	Hulled seed, creamy or pearly white, flattened, about 1/8-in. diam. Available: whole (unhulled and hulled).	Mildly nutty odor and taste.	14
FREEZE-DRIED SHALLOTS: California	1/4 x 3/8-in. white pieces.	Mild, but more aromatic onion odor and taste.	35
SWEET PEPPER FLAKES: California	1/4 x 1/4-in. flakes, bright green or red. Also granulated and powdered.	Mild, slightly sweet odor; somewhat bittersweet taste.	24
TARRAGON: California France Yugoslavia	As marketed, bits of green leaves. Available: whole and ground.	Sweet, aromatic, anise-like odor and taste.	50
THYME: Spain France	As marketed, bits of grayish to greenish-brown leaves. Available: whole and ground.	Fragrant, aromatic odor; warm, quite pungent taste.	20-1/2
TUMERIC: India Jamaica	Fibrous roots, orange-yellow in color; 1/3-in. long. Available: ground.	Characteristic odor, reminiscent of pepper; slightly bitter taste.	12

Soy on Menus

Use the following menu suggestions as ideas for your own operation. Choose one item from each line in each menu, and combine for a table d' hote price. Or list as given, with the comment on your menu that pricing is a la carte.

It is not to your advantage to offer more than two items containing soy protein in the same meal. If a customer chooses an all-soy meal, he may not appreciate the "flavor build-up." This seems to be a "funny feeling" at the back of the throat (it is really a subtle flavor which cannot be called by any of the usual sweet, sour, bitter or salt names). For people accustomed to eating soy protein, there is no problem. For those trying it for the first time, suggest a sweet or ice cream for dessert. Also, see "Cover Ups" (page 29).

BREAKFASTS

Hotel Room Service—Hearty Breakfast
Pear Halves   Kadota Figs   Green Gage Plums
Spanish Omelet*   Shirred Eggs
Cottage Fries   Potato Pancakes with Bacon-Flavor Bits
Brioche   Tavern Corn Muffins*   Toasted Date Fruit Bread*
Butter   Marmalade
Coffee

Coffee Shop or Dining Room or Restaurant—Specialty Breakfast
Cranberry Juice   Stewed Prunes   Orange and Grapefruit Sections
Sunnyside Eggs   Scrambled Eggs with Bacon-Flavor Bits
Fruit Kuchen*   Jelly Doughnuts
Milk   Cocoa   Cafe au Lait

Only for Conventions and Special Banquets
Breakfast to make getting up worth the effort
Bloody Mary   Screwdriver   Hard Cider   Pineapple Chunks in Gin
Poached Eggs on Toast   Eggs Rarebit*
Bran Muffins   Date Muffins   Blueberry Muffins
Coffee   Coffee with Chickory   Coffee with Brandy

Hospital or Nursing Home—Regular Diet Breakfast
Orange Juice   Apple Juice   Banana
Dry Cereal with Milk and Sugar
Grits with Butter or Milk and Sugar
Boiled Egg   Creamed Eggs
Bacon-Flavor Soy Protein Strips
Toast   Toasted English Muffin   Glazed Doughnut
Butter   Jelly
Coffee   Milk   Sanka   Tea

Coffee Shop or Cafeteria—Keep the grill hot and busy.
Baked Apple   Orange Juice   Peach Half
Scrambled Eggs   Fried Eggs
Scrapple*
Toast with Butter
Coffee   Milk

*Recipes included in this book—see Index

BRUNCHES

All-Purpose Buffet

Melon Slices   Fresh Fruit Cup   Stewed Raisins and Peaches   Orange Juice
Glazed Chicken Sandwiches*   Eggs Rarebit*   Creamed Eggs and Mushrooms
Chicken Livers with Chives in Sherry   Breakfast Steak
Hash Browns   French Fries
Corn Muffins*   Danish Pastry   Swirl Rolls*   Hot Biscuits   Pecan Rolls
Butter   Jelly   Marmalade
Coffee   Milk   Cocoa   Tea

Table d'Hote Brunch
Tangerine Juice
Eggs Benedict*
Croissant or Kugelhopf
Honey Butter
Coffee   Tea   Milk

White Table Cloth Brunch—To bring in the Sunday strollers
Champagne with Strawberries   Blueberries with Sour Cream and Brown Sugar
Hot Pepper Quiche*   Soofels* with Popadic's Spicy Tomato Sauce*
Toast   Popovers   Danish Pastry
Sweet Butter   Jam
Dessert Coffees*   Coffee   Spiced Tea   Tea

LUNCHES

A Regular Lunch for Office Workers
Beef Vegetable Soup*   Creamy Chicken Chowder*
Crackers
Finger Sandwiches   with   Pickles
Cola Soft Drink   Coffee

A Light Lunch with a Quick-Energy Boost
Consomme with Soy Strip Stirrer
Fancy Dress Sandwich*
Lemonade   Orangeade

Lunchtime with Kids at School
Chicken Noodle Soup*
Potato Chips
Celery and Carrot Strips
Peanut Butter and Jelly Sandwich
Peach half
Iced Tea   Fruit-Flavor Drink   Milk

The Ladies Luncheon for a Hotel Banquet
Cranberry Juice   Bloody Mary
Creamed Chicken* in Patty Shell
Dilled Green Bean Salad
Banana Bread   Whipped Cream Cheese
Iced Spiced Coffee

DINNER

Go ethnic sometime and make it frankly fake. Chinese food at a snack bar or in the coffee shop rather than the dining room is the idea.

* * * * * *

Fried Won Tons* Egg Rolls
Hot Mustard Sauce and Plum Sauce
Thin Sliced Cucumbers*
Chicken Oriental* Hamburger Chow Mein*
Steamed Rice
Fortune Cookies
Tea Beer Coffee

In a Family Dining Restaurant, Offer Adults a Home-style Meal
Stuffed Fruit* Gumbo Soup*
Sesame Crackers
Macaroni and Cheese
Broccoli
Sliced Tomatoes
Grasshopper Pie Lemon Chiffon Pie
Coffee Tea Milk Lemonade

In the Health-care Institution,
Offer a Treat Like Corn-on-the Cob for Regular Diets
Tomato Juice
Salisbury Steak with Gravy
Vegetables Vinaigrette (hot)* Corn-on-the-Cob
Fruit Cocktail in Orange Gelatin with Whipped Topping
Cookies
Milk Tea Coffee

MINI-MEALS

For Late Evening or that Too-Late-for-Lunch Hour
Spanish Omelet
French Bread with Sweet Butter
Sangria

Hot Pepper Quiche*
Caesar Salad
Tea

Cover-Ups

By itself soy protein isolate has a bland flavor; soy protein concentrate is also bland. Soy flour or soy grits may build up a bitter or "beany" flavor if eaten without other foods and flavors. Below are ideas and suggestions for using soy products and maintaining their pleasant taste.

Apples—all kinds (except green apples) are good with soy products, but additional apple juice or apple jelly is needed for a definite apple flavor.

Bananas—useful in desserts and specialty sandwiches with soy products. Also, if a snack or breakfast cereal is fortified with soy protein, banana flavoring or pieces of banana may be added for improved flavor.

Brown sugar—either brown or granulated sugar enhances many vegetable flavors, but with soy products brown sugar adds a rich mellow flavor note which is generally desirable. Even many entrees benefit from a bit of brown sugar added along with spices. Many of the bacon-flavor soy-base products available now contain sugar.

Butterscotch—very nice things happen with soy and butterscotch, especially in cookies and dessert flavors. But —a poor imitation butterscotch flavor may be worse than dreadful.

Carob powder—not very desirable because a synergism between soy and carob usually increases bitterness.

Chocolate—cocoa and unsweetened chocolate are "sometimes" flavorings to use with soy products. The lower fat cocoas and chocolate are less satisfactory with full-fat soy products and soy grits. However, addition of some other fats may yield an acceptable flavor combination. Milk chocolate, semi-sweet chocolate or a chocolate liqueur will be the best choice for use with all soy products. In short, soy products flavored with chocolate give a better flavor with added sugar and fat or milk solids (powdered dry milk).

Cider—a good choice for rehydrating dry soy protein because it gives an unusual and pleasant flavor when used in limited amounts with soy protein.

Cocktails—sweet cocktails with a fruit base such as mai tai or daiquiri give an unusual flavor to soy protein entrees. Use in limited amounts and do not marinate soy for longer than 15 to 30 minutes in the cocktails.

Coffee—"instant" or dried coffees change the slight bitterness of soy flours to an astringent quality which is definite but not unpleasant in a tea biscuit type cookie or cracker (only moderate amounts of sugar and fat). Brewed coffee (liquid) may be substituted for water or milk in some recipes using soy flour but the final flavor is usually better with 2 tsp. of vanilla for each 1 cup of brewed coffee.

Curry—especially if high in ginger and garlic, this spice blend permeates soy protein products and makes a new creation to delight neophytes as well as traditionalists among curry lovers. "Second day" curry is not recommended, but try braising soy protein for several hours in a curried beef, veal or chicken stock. Milk solids (powdered dry milk) make the sauce opaque, but add a better character to the final dish. All soy protein (no meat) can be used in a curry with milk solids for a widely acceptable product.

Dates—especially nice when a very few diced dates are added to a tart tomato sauce for use with soy protein. The date-filled cookie made with soy flour also has a pleasant flavor with wide appeal. Diced dates added to a dish with beef-flavored soy protein (beef stew, spaghetti sauce) make a more palatable entree. In using dates this way, consider them as a spice—no definable date flavor should be tasted, yet dates will round out the flavor of the entire dish.

Eggs—breakfast eggs can be happily combined with bacon-flavor soy protein. Omelets can be made with small additions of powdered soy protein isolate and milk, if a very high protein dish is needed. In using soy flour for breads and cookies, use eggs for emulsifying the dough and giving better texture and flavor to baked goods.

Fats—when possible, use the appropriate animal fat for meat-flavored soy protein: chicken fat for poultry-flavor items, bacon fat or lard for ham- or

More Cover-Ups

pork-flavor items. For beef-flavor items, use butter.

Margarines vary in composition of vegetable fats. Hydrogenated cottonseed oil and soy oil are the usual fats and are used in varying proportions. The solids/fat index determines "melt down," and in general the higher amounts of solids at higher temperatures give a better flavor. So a margarine which melts at about 95F is apt to be a good choice.

Better yet use a vegetable oil: coconut, cotton, peanut, corn, soy (in that order of preference, with coconut giving best flavor). Prices are in a curious state of flux and may remain so for some time; generally soy oil is least expensive.

Fish—definitely flavored fish, such as tuna and salmon, can be successfully used in casserole and salad recipes that are extended with unflavored soy protein or soy grits. FPC (fish protein concentrate) plus soy protein is a combination for snacks, baked products or pasta-type items which may be feasible on a commercial production basis.

Fruits—because most fruits are sweeter than vegetables, try using fruits with a variety of soy products. The fruits especially good for enhancing flavors of soy products are listed separately. Combine citrus fruits with peaches, cherries or pears for use with the soy products. Citrus juices, in small amounts, make a tasty sauce for soy entrees. Try pureed fruits as thickeners for sauces.

Garlic—although not always socially acceptable, a bit of garlic gives zest to soy protein entrees. (Rumor says the lemon pie filling served at Chicago's Art Institute is enhanced with a discreet amount of garlic, but since there is no reasonable way to work in any soy product, no tests were made on lemon meringue pie.)

Honey—somewhat more flavorful on its own than brown sugar, honey covers the flavor sins of soy flour very well. Special techniques (such as adding the honey to mixtures with eggs or with fat during recipe preparation) are easy to adapt. Baked goods and desserts made with soy products taste better if honey is used as a sweetener.

Jams, Jellies, Preserves—like pureed fruits, these sweets are pleasing thickeners for sauces. Make a Cumberland Sauce of 2 parts brown stock or au jus, 1/2 part sherry, 1/2 part jam. Simmer and pour over heated soy protein.

Nuts—finely chopped nuts or nut meal will enrich soy product recipes. Entrees and sandwich filling with ground soy and ground nuts are tasty. Peanuts and pecans add the most pleasant flavors, although English walnuts are acceptable.

Oats—unflavored soy grits or soy protein isolate can be combined with quick cooking oats for crumb-type toppings or as extenders in ground meat products. In a multi-cereal bread, such as a "healthful" loaf of bread made with soy and wheat flours, oats smooth out competing flavors to give mellowness to entire mixture.

Olives—ripe olives are a better choice for use with soy products because they are less bitter or astringent than the green olives.

Onions—whether dried flakes, yellow or white "regular" onions or green onions (scallions), the onion is like garlic. It takes your taste sensor in tow down a strong-flavored path which permits little attention to any drawbacks of soy products' flavors. However, in some breads and sauces, toasting the dried onion flakes before rehydrating eliminates their "dehydrated" flavor.

Peaches—canned peaches are a good combination with soy products.

Peanut Butter—for desserts and baked products with soy flour, peanut butter with honey or brown sugar seems to have a natural affinity. The peanut flavor is less strident and the soy flavor never appears as that "catch in the back of the throat."

Soy Sauce—go easy with this, even in Oriental dishes. Try a bit of bead molasses or salt and MSG instead.

Wheat Germ—offers similar advantages when used with or instead of oats in soy product recipes.

Soy Breakfast Items

Strictly convenience—ADM Smoky Bits (TVP chips) are bacon-flavor soy protein pieces just made for topping off fluffy scrambled eggs.

Archer-Daniels-Midland

More Soy Breakfast Items

EGGS BENEDICT
YIELD: 8 servings
INGREDIENTS:

Bontrae frozen Ham-Flavor Soy Protein chunks	8 oz.
Land O'Lakes Butter	as needed
Hollandaise sauce	1 pt.
English muffins, split, toasted and buttered	4
Poached eggs	8

METHOD: Heat ham quickly in skillet with small amount of butter. Stir into hot Hollandaise sauce. Serve over poached eggs on muffins.

EGGS RAREBIT
YIELD: 8 servings
INGREDIENTS:

Canned cheese sauce	1 pt.
Beer	1/2 cup
Bontrae frozen Ham-Flavor Soy Protein chunks, thawed	8 oz.
Worcestershire sauce	1 tsp.
Tabasco	3 drops
Chopped chives	1 Tbsp.
English muffins, split and toasted	4
Poached eggs	8

METHOD: Heat cheese sauce and beer in saucepot, stirring until smooth and blended. Add Bontrae, Worcestershire sauce, Tabasco and chives; simmer 15 minutes.

On each muffin half place poached egg; spoon sauce over top.

SCRAPPLE
YIELD: 20 (1/2-in.) slices
INGREDIENTS:

Ground beef	8 oz.
Quaker Instant grits	1 lb.
Bontrae frozen Beef-Flavor Soy Protein crumbles, thawed	6 oz.
Durkee Flavor salt	1 tsp.
Tabasco	1/4 tsp.
Boiling Water	2 qt.

METHOD: Brown ground beef in large skillet or pot. Remove from heat.

Combine ground beef with grits, Bontrae, salt and Tabasco. Stir boiling water into grits mixture. Continue stirring until thoroughly mixed.

Pack mixture into oiled 9- by 5- by 3-in. bread pan. Chill until firm. Slice 1/2-in. thick. Dust with flour if desired. Fry in hot fat.

SERVE Scrapple with baked apples and fruit-gelatin salad for lunch. OR Serve platters of scrambled eggs with Scrapple slices and garnish of skewered orange sections and plumped pitted prunes or dates.

SPANISH OMELET
YIELD: 2 servings
INGREDIENTS:

Bontrae frozen Ham-Flavor Soy Protein chunks	1/2 cup
Onion, sliced	1 medium
Land O' Lakes Butter	1/4 cup
Tomato pieces with juice	1-1/2 cups
Dehydrated diced green pepper	2 Tbsp.
Kleer Gel (thickener)	1 Tbsp.
Water	1/2 cup
Fresh whole eggs	1 cup (5-6 eggs)
Salt	1/2 tsp.

METHOD: In saucepan saute Bontrae and onion in butter for 2-3 minutes. Add tomatoes with juice and green pepper. Make a smooth paste of thickener and water. Stir into tomato mixture. Cook and stir just until mixture is thickened or has come to a boil.

Separate eggs. Add salt to egg whites; whip until soft peaks form. Whip egg yolks until fluffy. Gently fold egg yolks into whites. Turn egg mixture into hot oiled omelet pan or skillet. Cook over medium heat about 5 minutes, until omelet is puffy. Cover with tight lid. Cook about 5 minutes longer, or until top of omelet is set.

Slide omelet onto serving platter; spoon on half of tomato sauce. Fold omelet in half. Spoon remaining sauce over top.

Breakfast Put-Ons

The syrup or toppings put on waffles, pancakes and French toast are often more important than the waffle.

Here are some ideas for "put-ons" for breakfast as well as luncheon waffles and related breadstuffs.

SQUARE PANCAKES: Prepare a 5 lb. package of pancake mix as directed. Pour into a greased 18- by 26-inch bun pan. Bake at 375F for 15-20 minutes. Cut into squares and serve hot.
    Variations--stir 8 oz. bacon-flavor soy protein into pancake batter before panning.
        --saute 5 oz. dry unflavored fine-minced soy protein in 8 oz. butter until lightly browned. Stir in 1/2 cup sugar. Add this mixture to pancake batter before panning.
        --mix 1/2 cup soy flour with 1 cup wheat germ and 1 tsp. cinnamon. Sprinkle over top of pancake batter in pan before baking.

CREAMED EGGS: Serve creamed eggs over waffles. Garnish with bacon flavor soy bits.

GARNISHED DEVILED EGG: Garnish a deviled egg half by wrapping in a fried Stripple (bacon-flavor soy protein strip). Secure with wooden pick. Serve with cheese sauce over waffles.

WAFFLE BLT: Make a BLT sandwich using waffles instead of bread, 1000 Island Dressing served "on the side" instead of plain mayonnaise, and Stripples instead of bacon.

HAM & CHEESE: Use pre-prepared French toast as a base for an open-face sandwich topped with ham-flavor soy protein in cheese sauce. Garnish with sweet relish.

SWEET BBQ: Serve Barbecue Ham and Cherries in Sauce (recipe on page 43) over French toast.

More Breakfast Put-Ons

Relishes are great for waffles and and pancakes. Try some of these:

CHERRY CHUTNEY--add 1 Tbsp. ground cinnamon and 2 Tbsp. vinegar to a No. 10 can cherry pie filling.

CORN RELISH--serve grilled Vega-Links (all-soy hot dogs) with corn relish on pancakes.

CRANBERRY-ORANGE RELISH-- serve with a version of Square Pancakes.

CHILI SAUCE--use with creamed chipped beef on crisp waffles.

SPICED PURPLE PLUMS–cook 1-1/2 lb. dried prunes in 2 qt. water with 2 sticks cinnamon, 10 whole cloves and 1 Tbsp. each brown sugar and vinegar for 45 minutes. Serve with the Garnished Deviled Egg (recipe on page 33).

QUEEN'S BLT--prepare as directed (recipe on page 33) omitting 1000 Island Dressing. Beat 1 egg white until soft peaks form. Gently fold in 2 Tbsp. mayonnaise. Spoon onto 1 waffle square. Broil just until puffy and golden. Use as top of BLT.

SPICY RAISINS--cook 1-3/4 lb. raisins in 2 qts. water until raisins are plump. Stir in 1 cup apple butter, 1/4 cup sugar, 1/4 cup vinegar, 2 Tbsp. prepared mustard. Add 1 lb. frozen ham-flavor soy protein chunks. Simmer 15 minutes. Serve on pancakes, French toast.

MUSTARD SAUCE--blend together 1 qt. mayonnaise with 1-1/2 cups prepared mustard. Gently fold into 1 qt. prepared whipped topping. Serve on pancakes; top with bacon flavor bits.

CUSTARD SAUCE-- serve over French toast. Top with broiled peach half.

HOT FUDGE SUNDAE--make a waffle sundae: scoop of vanilla ice cream on a waffle. Top with hot fudge sauce.

APPLESAUCE--heat applesauce. Stir 1 tsp. ground cinnamon into each cup used. Serve over toasted Date Fruit Bread.

MARINATED LENTILS

YIELD: about 2 qt.

INGREDIENTS:

Lentils	2 lb.
Dehydrated onion flakes	1 cup
Water	3 qt.
Prepared Italian salad dressing	1 pt.
Salt	1 Tbsp.
Garlic Powder	1/4 tsp.
Prosage Sausage-Flavor Soy Protein	1 lb.

METHOD: Cook lentils and onion flakes in water just until lentils are tender—about 20-30 minutes. Drain. In large bowl combine cooked lentils and onion flakes with dressing, salt and garlic powder. Toss to mix well. Cover. Let marinate in refrigerator at least 24 hours.

Just before serving, thinly slice or dice Prosage. Fry in deep fat at 375F for about 2 minutes, until lightly browned. Mix with lentils. Serve at once. Garnish with green olives, if desired.

PATÉ A L'ORANGE

YIELD: 100 servings, about 2 oz. each

INGREDIENTS:

Orange flavor gelatin	12 oz.
Boiling water	3 pt.
Mandarin oranges, drained	1 qt. (approx.)
Bontrae frozen Ham-Flavor Soy Protein, thawed, ground	5 lb.
Mayonnaise	2 qt.
Prepared mustard	1 cup

METHOD: Dissolve gelatin in boiling water. Pour 1 cup into each of 5 bread pans, 5- by 9- by 3-in. Place a single layer of orange sections in gelatin in pans. Chill.

Combine Bontrae, mayonnaise and mustard in mixer bowl. Mix on low speed, adding remaining gelatin slowly. Continue mixing until well blended. Divide into pans. Spread evenly in each pan, packing gently. Chill until firm. Slice about 1/2-in. thick. Serve 1 slice on lettuce leaf with green olives and crackers or Melba toast.

Soy Appetizers-First Course Items

S and W Fine Foods

Little Sausage balls for Hibachi service can be made from Meat Loaf I or Meat Loaf II (page 42).

More Soy Appetizers

VEGETABLES VINAIGRETTE
YIELD: 48 servings
INGREDIENTS:

Green beans, drained	1 No. 10 can
Small white potatoes, drained	1 No. 10 can
Salad oil	1 cup
ReaLemon Lemon juice	3/4 cup
Paprika	1 tsp.
Salt	2 tsp.
Pepper	1/2 tsp.
Garlic powder	1/4 tsp.
Ground sage	1/4 tsp.
ADM Smoky Bits Bacon-Flavor Soy Protein	8 oz.

METHOD: In saucepot or steamjacketed kettle, combine vegetables, oil, lemon juice and spices. Cook on low heat for 15-20 minutes, until vegetables are heated through.

To serve hot, toss with Smoky Bits and serve at once.

To serve cold, chill at least 6 hours. Toss vegetable mixture with smoky bits just before serving. Serve on lettuce and garnish with chopped pimiento, if desired.

FRIED WON TONS
YIELD: 24 appetizers
INGREDIENTS:

Bontrae frozen Ham-Flavor Soy Protein chunks, thawed	1 lb.
Cooked ham	4 oz.
Canned tartar sauce	3/4 cup
Pie crust dough	2 lb.

METHOD: Grind together Bontrae and ham. Mix in tartar sauce. Set aside. Roll out dough to form 2 rectangles, 12- by 24-in. each. Cut dough into squares, 3- by 4-in. each. Place 1 oz. ham mixture in center of each square. Pinch dough together to seal around filling, leaving ends open.

Fry in deep fat at 375F for 2 to 3 minutes, until golden. Serve hot.

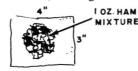 

STUFFED FRUIT
YIELD: about 100 pieces
INGREDIENTS:

Cream Cheese	1 lb.
Frozen orange juice concentrate, thawed	1/2 cup
Sherry	1/4 cup
Bordo Pitted dates OR Pitted prunes	1-1/2 lb.
Bacon-Flavor Soy Protein pieces	as needed

METHOD: Beat cream cheese just to soften. Add orange juice concentrate and sherry; beat just until blended. Fill cavities of fruit with cream cheese mixture, using heaping teaspoonful for each. Roll filled side of fruit in bacon-flavored soy so several pieces adhere to cream cheese filling. Or, place 2 or 3 pieces of bacon-flavor soy on filling of each piece of fruit.

NOTE: For best results, do not mix soy pieces into cream cheese filling.

APPETIZER ROLL
YIELD: About 48 servings
INGREDIENTS:

Ground beef	6 oz.
Bontrae frozen Hamburger-Flavor Soy Protein crumbles	6 oz.
Condensed cream of celery soup, undiluted	2-1/2 cups
Dried onion flakes	1/4 cup
Dehydrated diced green peppers	1/4 cup
Durkee Flavor salt	1 tsp.
Tabasco	1/2 tsp.
Pastry dough	3 lb.
Cheese sauce or mushroom sauce	as needed

METHOD: Brown ground beef; drain off fat and combine meat with remaining ingredients. Mix well. Divide pastry dough into three equal portions; roll each into a 12- by 15-in. rectangle.

Spread one-third ground beef mixture over each rectangle (1-2/3 cups per roll). Roll up each dough rectangle, starting with 15-in. edge, jellyroll fashion. Place, cut-side down, on greased 18- by 26-in. pan.

Bake at 425F in a conventional oven for 25 to 30 minutes; OR Bake at 350F in a convection oven for 15 to 20 minutes. Cut into 1/2-in. slices and serve with cheese or mushroom sauce.

Olive Administration Committee

Plate garnishes can be plain, fancy, large or small. All make foods more fun to eat. Clockwise from the Lower Left: pitted ripe olives stuffed with tiny shrimp; cucumber slice twisted so cherry tomato is skewered on one side and ripe olive is on the other side; kebab of ripe olive-pickled artichoke heart-cherry tomato-ripe olive; kebab of thin sliced salami twisted and skewered between ripe olives; sprigs of watercress hold sliced ripe olives; lettuce cup with tomato slice sprinkled with chopped onion and ripe olives then seasoned with salt and a dash of vinegar or freshly ground black pepper; citrus slices (they may be orange, lemon or lime) have long slices of ripe olives arranged petal fashion.

IDEAS:
The tomato garnish or salami-olive kebab dress up the appetizer plate. Try one with the Marinated Lentils or Vegetables Vinaigrette.
Skewer a shrimp-stuffed olive on top of a chicken salad sandwich, then add an olive-decorated sprig of watercress.
Paté a L'Orange becomes more festive with a citrus cartwheel.

DATE PIZZA
YIELD: 2 (18- by 26-in.) pizzas, 468 hors d'oeuvre servings
INGREDIENTS:
Pillsbury Pizza crust mix	2 lb. 8 oz.
Water	as required
Bordo Diced dates	5 lb.
Heinz Pureed peaches	2 (16 oz.) cans
Orange juice	1-1/3 cup
Water	1 qt.
Sherry	1 cup
Bacon-Flavor Soy Protein	
OR	
ADM Large Smoky Bits	8 oz.

METHOD: Prepare pizza crust mix with water as directed on package. Pat half of dough into each of 2 greased 18- by 26-in. pans. Bake crust at 450F in a conventional oven for 10 minutes;
OR
Bake at 350F in a convection oven for 5 minutes.

Combine dates, peaches, water, orange juice and sherry; cook in saucepot until thickened. Remove from heat; stir in bacon bits. Spread date mixture over partially baked pizza crusts.

Bake at 450F in a conventional oven for 20 minutes;
OR
Bake at 350F in a convection oven for 12 minutes. Cut into 1- by 2-in. pieces.

APPETIZER CHUNKS
YIELD: About 2 qt.
INGREDIENTS:
Honey	1 pt.
ReaLime Lime juice	1/3 cup
Nutmeg	1 tsp.
Cloves	1/4 tsp.
Bontrae frozen Ham-Flavor or Chicken-Flavor Soy Protein chunks, thawed	2 lb.

METHOD: Combine honey, lime juice and spices. Simmer about 5 minutes. Add Bontrae, simmer 5 minutes longer. Serve in chafing dish as appetizer with cocktail picks.

Soy Soups

CREAM OF POTATO SOUP
YIELD: 50 (5 oz.) servings
INGREDIENTS:

Continental Dehydrated hash brown potatoes	10 oz.
Dehydrated onion flakes	2 cups
Water	3 qt.
Instant mashed potatoes	2 cups
Powdered dry milk	6 cups
Milk	2 qt.
ADM Smoky Bits or Bacon-Flavor Soy Protein bits	6 oz.
Salt	4 tsp.
Dried parsley flakes	1/4 cup

METHOD: Rehydrate hash brown potatoes and onions in water in steam-jacketed kettle. Add dry mashed potatoes and dry milk slowly, stirring to dissolve. Add remaining ingredients.

Simmer about 20 minutes, stirring occasionally, until heated through.

CHICKEN NOODLE SOUP
YIELD: 48 (6 oz.) servings
INGREDIENTS:

Water	2 gal.
Whole carrots, peeled	6
Whole onions, peeled	3
Rendered chicken fat or Land O'Lakes Butter	4 oz.
Chicken soup base	1/2 cup
Noodles, uncooked	1 lb.
Bontrae frozen Chicken-Flavor Soy Protein chunks*	2 lb.
Cooked chicken, diced*	1 lb. 4 oz.

METHOD: Combine all ingredients except soy protein chunks, cooked chicken and noodles in saucepot or steam-jacketed kettle. Cook 30 minutes. Add remaining ingredients. Bring to boiling; reduce heat and simmer 15-20 minutes or until noodles are done. Serve hot.
NOTE: Carrots and onions are for flavoring only and may be discarded before serving.

*AMOUNTS FOR SCHOOL LUNCH: Use 15 oz. frozen Bontrae and 2 lb. 4 oz. cooked chicken. Has 1 oz. protein per serving.

PEA SOUP
YIELD: About 1 gal.
INGREDIENTS:

Warm water	3 qts.
ADM dry Ham-Flavor Soy Protein chunks	8 oz.
Dehydrated onion flakes	2 Tbsp.
Condensed pea soup, undiluted	1 (50-oz.) can
Worcestershire sauce (optional)	1 Tbsp.

METHOD: In saucepot or steamjacketed kettle, combine water, dry soy protein and onion flakes. Let rehydrate 15 minutes. Stir in pea soup and Worcestershire sauce. Bring to boiling; reduce heat and simmer 10-15 minutes.

FOR GREATER FLAIR: Substitute 3 cups Bontrae frozen ham-flavor soy protein chunks for dry soy protein. Reduce water to 3 qts. Garnish each serving of soup with a dollop of sour cream and bacon-flavor soy protein bits.

GUMBO
YIELD: 25 (8-oz.) servings
INGREDIENTS:

Water	7 cups
Tomato juice	2 (46-oz.) cans
Tomatoes, undrained, broken	1 No. 2-1/2 can
Sliced okra, drained	2(1 lb.) cans
Canned diced celery, undrained	1 lb.
Canned diced onions, undrained	1 lb.
Bontrae frozen Chicken-Flavor Soy Protein chunks	1 lb.
Rendered chicken fat	6 oz.
Dehydrated diced green pepper	3 oz.
Chicken soup base	1 tsp.

METHOD: Combine all ingredients in steamjacketed kettle. Mix well. Cook 45 minutes. Serve as soup or as sauce over cooked rice.

COST BONUS: Rehydrate 8 oz. dry soy protein granules in 1-1/2 cups of water and substitute for chicken soy protein.

BEEF VEGETABLE SOUP

YIELD: about 4 gal., 85 (6-oz.) servings

INGREDIENTS

Water	2 gal.
Tomatoes	1 No. 2-1/2 can
Peas and onions, undrained	1 No. 10 can
Barley	1 lb.
Durkee dried mixed vegetable flakes	1 pt.
Bouillon cubes	4
Brown sugar	3 oz.
ReaLemon Lemon juice	3 Tbsp.
Durkee Flavor salt	1 Tbsp.
Cooked beef, cubed*	1 lb. 8 oz.
Worthington dry Hamburger-Flavor Soy Protein granules*	1 lb.

METHOD: Combine all ingredients except cooked beef and dry soy protein in saucepot or steamjacketed kettle. Simmer 1 hour. Add soy protein granules and cooked beef. Simmer 20 minutes.

COST BONUS: Substitute 1 (4 lb. 12 oz.) can beef-flavor soy protein chunks, well drained, for dry soy protein granules plus cooked beef.

*AMOUNTS FOR SCHOOL LUNCH: Use 10 oz. dry soy protein granules and 4 lb. cooked beef. Has 1 oz. protein per serving.

CORN CHOWDER

YIELD: About 2 gal.

INGREDIENTS:

Whole kernel corn, drained	1 No. 10 can
Milk	3 qt.
Bontrae frozen Ham-Flavor Soy Protein chunks	2 lb.
Land O' Lakes Butter	4 oz.
Chopped celery	6 oz.
Chopped onion	6 oz.
Salt	1 Tbsp.
Pepper	1 tsp.
Water	1 pt.
Dehydrated yam flakes	10 oz.

METHOD: Combine all ingredients in steamjacketed kettle, except water and yam flakes. Then dissolve yam flakes in water. Stir into chowder. Cook 30 minutes, stirring occasionally.

CREAMY CHICKEN CHOWDER

YIELD: About 9 qt., 50 (6-oz.) servings

INGREDIENTS:

Condensed cream of celery soup, undiluted	2 (50 oz.) cans
Milk	2 qt.
Water	2 qt.
Cream-style corn	2 (1 lb.) cans or 1 qt.
Dried parsley flakes	2 Tbsp.
Bontrae frozen Chicken-Flavor Soy Protein chunks, thawed*	1 lb. 12 oz.
Cooked chicken, cubed*	1 lb. 8 oz.

METHOD: In saucepot or steamjacketed kettle, combine all ingredients except Bontrae and chicken. Stir to mix. Simmer 15 minutes. Add Bontrae and chicken. Simmer 15 minutes in saucepot or cook 10 minutes in steamjacketed kettle.

*AMOUNTS FOR SCHOOL LUNCH: Use 1 lb. frozen Bontrae and 2 lb. 4 oz. cooked chicken. Has 1 oz. protein per serving.

CHILI SOUP

YIELD: about 2-1/2 gal., 40 (1-cup) servings

INGREDIENTS:

Ground beef*	2 lb.
Bontrae frozen Beef-Flavor Soy Protein crumbles*	1 lb. 12 oz.
Chili beans	1 No. 10 can
Water	1 gal.
Tomato sauce	1 (29-oz.) can
Dried onion flakes	1 cup
Dehydrated diced green pepper	1/3 cup
Salt	2 tsp.

METHOD: In large pot or steamjacketed kettle, brown ground beef; add Bontrae and stir until well mixed. Add remaining ingredients. Stir to mix thoroughly. Cook over low heat 20 to 30 minutes, stirring occasionally—do not boil. Serve with crackers.

*AMOUNTS FOR SCHOOL LUNCH: Use 12 oz. frozen Bontrae and 2 lb. 8 oz. ground beef. Has 1-1/2 oz. protein per serving.

Garnishes for Soy Soups

Libbey Products, Owens-Illinois

Serve soup in a mug. Use clear soups in glass mugs and footed mugs for chowders. Add a "stirrer" of bacon-flavor soy protein (Stripples).

1. Float bacon-flavor soy protein pieces on cream of mushroom soup.

2. Stir-fry dry hamburger or beef-flavor soy protein in butter. Sprinkle with garlic salt. Serve in cream of potato soup or tomato soup.

3. Put dry un-flavored soy protein chunks in ham or brown stock. Let soak (heated or unheated) about 1 hour. Add to bean soup (use 1 or 2 Tbsp. per serving).

4. Use dry ham-flavor soy protein pieces as garnish for tomato bouillon.

5. Thinly slice sausage-type soy protein. Deep fry at 375F for 1-1/2 minutes. Place 1 or 2 slices on lentil soup or corn chowder. Also nice on French onion soup with grated Parmesan.

6. Toast dry soy protein chunks (any meat flavor or plain) and whole shelled almonds. Use shallow pan filled 1 layer deep. Roast at 450F for 2-3 minutes. Use for garnish on cream soup or consomme.

7. Make small meat balls of Meat Ball Hoagie mixture (page 60). Brown or deep fry and use as garnish for clear soups such as consomme or bouillon. (see picture, page 45.)

Soy Entrees-Side Dishes

Crisp taco shells or soft tortillas can be served with Burrito Filling (see Taco Filling I recipe variation, page 43) and a variety of garnishes. Use larger amounts for entree, less for appetizer. Include chopped onions and tomatoes, grated cheese, sour cream, guacamole, shredded lettuce. Then let the do-it-yourself fans help themsleves. Good on small buffets too.

ReaLemon

More Soy Entrees

MEAT LOAF I
YIELD: 5 lb. loaf
INGREDIENTS:

Worthington dry Hamburger-Flavor Soy Protein granules*	8 oz.
Dehydrated onion flakes	3/4 cup
Dried parsley flakes	1/4 cup
Dehydrated diced green peppers	1/4 cup
Warm water	5 cups
Ground beef*	3 lb.
Bread crumbs	1 pt.
Worcestershire sauce	2 Tbsp.
Salt	4 tsp.
Pepper	2 tsp.

METHOD: In large mixing bowl, rehydrate soy protein, onions, parsley and green pepper in warm water for 15 minutes. By hand or with paddle on low speed of mixer, blend together all ingredients.

Place in 12- by 20- by 4-in. steamtable pan. Shape into 2 loaves, the length of the pan. Bake at 350F for about 1 hour and 15 minutes.

*AMOUNTS FOR SCHOOL LUNCH: Use 6 oz. dry soy protein and 3 lb. 8 oz. ground beef. Has 2 oz. protein per serving, 28 cuts per 5 lb. loaf.

MEAT LOAF II
YIELD: 5 lb. loaf
INGREDIENTS:

Worthington dry Hamburger-Flavor Soy Protein granules*	1 lb.
Warm water	7 cups
Ground beef*	3 lb. 6 oz.
Bread crumbs	1 pt.
Catsup	1 cup
Eggs	2

METHOD: Rehydrate dry soy protein in warm water for 15 minutes in mixer bowl. Add remaining ingredients. Mix with paddle on mixer at low speed. Shape into loaf in roasting pan. Bake at 350F for 1 hour. Serve with mushroom sauce or catsup.

*AMOUNTS FOR SCHOOL LUNCH: Use 6 oz. dry soy protein and 3 lb. 8 oz. ground beef; reduce water to 5 cups. Has 2 oz. protein per serving, 30 cuts per 5 lb. loaf.

CREAMED CHICKEN
YIELD: 25 (4-oz.) servings
INGREDIENTS:

Cooked chicken, cubed	1 lb. 12 oz.
Bontrae frozen Chicken-Flavor Soy Protein chunks, thawed	1 lb. 8 oz.
Condensed celery soup, undiluted	1/2 (50-oz.) can
B & B mushrooms, undrained	2 (6-oz.) cans
Chopped pimientos	2 oz.

METHOD: Combine all ingredients in saucepot. Simmer 15 minutes, stirring occasionally.
COST BONUS: Substitute 3 (13-oz.) cans Worthington Soya chicken style, drained, for the frozen Bontrae.

HAM TETRAZZINI
YIELD: 25 servings
INGREDIENTS:

Spaghetti, uncooked	1 lb.
Boiling water	1 gal.
Condensed cream of celery soup, undiluted	1 (50-oz.) can
Milk	2 cups
Dried parsley flakes	2 Tbsp.
Salt	3/4 tsp.
Bontrae frozen Ham-Flavor Soy Protein chunks	1 lb.
Cooked ham, cubed	3 lb. 2 oz.
Parmesan cheese, grated	3 oz.
Paprika	1 Tbsp.

METHOD: Cook spaghetti in boiling water. Drain and rinse. Blend soup, milk, parsley, and salt. Combine Bontrae, ham and spaghetti in 12- by 20- by 4-in. steamtable pan. Pour soup mixture over top; mix thoroughly. Mix together cheese and paprika. Sprinkle top of spaghetti mixture with cheese and paprika. Bake at 400F in conventional oven for 30-40 minutes; OR
Bake at 325F in a convection oven for about 25 minutes.

NOTE: 1 lb. 10 oz. regular raw rice cooked in 6-1/4 cups water may be substituted for the cooked spaghetti.

TACO FILLING I

YIELD: About 3 qt.

INGREDIENTS:

Ground beef	1 lb.
Bontrae frozen Beef-Flavor Soy Protein crumbles, thawed	1 lb. 12 oz.
Tomato pieces with juice	1 pt.
Brown gravy	1 cup
Chili powder	1 Tbsp.
Coriander	1/2 tsp.
Ground cumin	1/2 tsp.

METHOD: Brown ground beef in saucepot or steamjacketed kettle. Add remaining ingredients and stir until well mixed. Cook 20-30 minutes.
VARIATION: For Burrito filling, add 1 qt. drained kidney beans and 2 Tbsp. ReaLemon Lemon juice to browned meat. If desired, mash filling before filling tortillas for the Burritos.

BARBECUE HAM AND CHERRIES IN SAUCE

YIELD: about 2-1/2 gal.

INGREDIENTS:

Dark sweet pitted cherries, undrained	1 No. 10 can
Red tart cherries, undrained	1 No. 303 can
ReaLime Lime juice	1 cup
Water	as needed
Sugar	1-1/3 cups
Dehydrated onion flakes	1/2 cup
Durkee Barbecue spice	1/2 cup
Cornstarch	1/2 cup
Bontrae frozen Ham-Flavor Soy Protein chunks	5 lb.
Cubed cooked ham	3 lb.

METHOD: Drain sweet and tart cherries; set cherries aside. Combine cherry juice with lime juice and enough water to measure 3 pints. In saucepot or steamjacketed kettle, combine measured liquid with sugar, onion flakes and barbecue spice. Make a smooth paste of cornstarch and water; add to juice mixture, stirring until well blended. Cook and stir until thick and clear. Add cherries, Bontrae and ham. Cook 30 minutes;
OR
Pour into 20- by 12- by 4-in. steamtable pan. Bake at 375F for 30 minutes. Serve hot with rice.

TACO FILLING II

YIELD: About 3 qt.

INGREDIENTS:

Ground beef *	1 lb.
Bontrae frozen Beef-Flavor Soy Protein crumbles, thawed	1 lb. 12 oz.
Tomato puree	1 pt.
Water	1 cup
Dehydrated onion flakes	1/2 cup
Chili powder	1 Tbsp.
Salt	1 tsp.
Ground cumin	3/4 tsp.
Garlic powder	1/4 tsp.

METHOD: Brown ground beef in saucepot. Add remaining ingredients and stir until well mixed. Cook until thickened, about 25 minutes.
SERVING TACOS: Fry the canned or frozen tortillas in deep fat at 375F for about 1-1/2 minutes, holding opposite sides together so tortilla forms a half-circle,
OR
Use pre-prepared taco shells. Force edges apart to fill.

For each Taco, use:

Taco Filling I or II	1 to 2 oz.
Grated cheddar OR Monterey Jack cheese	1 Tbsp.
Shredded lettuce	as needed
Chopped onion	1 Tbsp.
Chopped tomato	1 Tbsp.

Fill taco shell first with hot Taco Filling, then cheese and remaining ingredients.
*AMOUNTS FOR SCHOOL LUNCH: Each 1-1/2 oz. filling contains 3/4 oz. protein.

FRUIT SAUCE FOR SLICED LOAF

YIELD: 25 servings (4 oz. sauce with 2-1/2 oz. Wham)

INGREDIENTS:

Fruit cocktail with syrup	1 No. 10 can
California seedless raisins	6 oz.
Chopped chives	2 Tbsp.
Durkee Flavor salt	1 tsp.
Red currant jelly	10 oz. jar
Wham frozen Ham-Flavor Soy Protein loaf, thawed and sliced	4 lb.

METHOD: Combine fruit cocktail, raisins, chives and salt in saucepot. Cook until slightly thickened. Add jelly and stir until melted. Serve over hot sliced Wham.

More Soy Entrees

MEATLESS SPAGHETTI SAUCE

YIELD: About 3 qt.

INGREDIENTS:

Sliced onions	6 oz.
Land O'Lakes Butter	4 oz.
Worthington dry Hamburger Flavor Soy Protein granules	10 oz.
Warm water	3 pt.
ReaLemon Lemon juice	1 Tbsp.
Oregano leaves	2 tsp.
Sweet basil leaves	1 tsp.
Tabasco	1/4 tsp.
Garlic powder	1/8 tsp.
Bay leaf	1
Tomato puree	1 (29-oz.) can

METHOD: Saute onions in butter in steamjacketed kettle or saucepan. Turn off heat. Add dry soy protein and warm water. Let stand 20 minutes. Add remaining ingredients. Simmer 40 minutes.

FOR GREATER FLAIR: Cube and deep fry until browned 1 lb. Prosage soy protein. Combine with 8 oz. frozen ham-flavor soy protein chunks. Substitute for the dry hamburger granules. Reduce water to 1 qt.

WHISKEY YAMS WITH PEACHES 'n HAM

YIELD: 24 servings

INGREDIENTS:

Bontrae frozen Ham-Flavor Soy Protein chunks, thawed	1 lb. 12 oz.
Cooked ham, cubed	1 lb. 8 oz.
Cut yams, drained	1 No. 10 can
Cling peach halves, drained	2 No. 2-1/2 cans
Brown sugar	8 oz.
Whiskey	1 cup
ReaLemon Lemon juice	1/2 cup

METHOD: In 20- by 12- by 4-in. steamtable pan combine Bontrae and ham with yams and peaches. Make a paste of brown sugar, whiskey and lemon juice; spoon over yam mixture. Bake at 350F for 30 minutes.

Serve with a mild green vegetable such as buttered peas or green beans, corn bread, lots of iced tea, and suggest Lemon Meringue Pie.

AUDREY'S CHILI

YIELD: 50 servings

INGREDIENTS:

Worthington dry Hamburger-Flavor Soy Protein granules*	1 lb. 8 oz.
Warm water*	3 pts.
Ground beef*	6 lb.
Kidney beans, undrained	1 No. 10 can
Chili beans, undrained	1 No. 10 can
Tomato puree	8 (29-oz.) cans
Dehydrated onion flakes	2 cups
Salt	1/2 cup
Chili powder	1/3 cup
Sugar	1/3 cup
Pepper	2 tsp.
Vinegar	1 cup
Green olives stuffed with pimiento, drained	3 cups

METHOD: Rehydrate dry soy protein in warm water 30 minutes. Brown ground beef. Combine with remaining ingredients except olives. Cook in steamjacketed kettle 40 minutes. Stir in olives. Cook 15 to 20 minutes longer.

OR

Turn into two 12- by 20- by 4-in. pans. Bake at 375F for 45 minutes. Stir in olives. Cook or bake 15 minutes longer.

*AMOUNTS FOR SCHOOL LUNCH: Use 12 oz. dry soy protein, reduce water to 3 cups, use 6 lb. 4 oz. ground beef. With beans, 3 oz. protein per serving; without beans, 2 oz. protein per serving.

BEEF STEW

YIELD: 24 servings

INGREDIENTS:

ADM dry Beef-Flavor Soy Protein chunks	1 lb.
Cider	3 cups
Beef stew	1 No. 10 can
Peas	1 qt.
Kitchen Bouquet	3 Tbsp.
Oregano leaves	1/2 tsp.

METHOD: In 12- by 20- by 4-in. steamtable pan, rehydrate dry soy protein in cider 20 minutes. Add remaining ingredients; stir to mix. Bake at 375F, covered, for 40 minutes.

BEANY BAKE

YIELD: 21 servings

INGREDIENTS:

Pork and beans	1 No. 10 can
Catsup	1 (14 oz.) bottle
Brown sugar	1 lb.
Bacon-flavor soy protein bits	(optional)
Vega-Links All-Soy Hot Dogs, drained	1 (4 lb., 12 oz.) can

METHOD: Combine beans with catsup and brown sugar. Sprinkle with soy bits. Arrange hot dogs on top. Bake at 375F for 20-30 minutes.

FOR GREATER FLAIR: Omit bacon soy bits. Spread 1 pt. canned cheese sauce over top of hot dogs 15 minutes before removing from oven.

CHICKEN ORIENTAL

YIELD: 25 (1 cup) servings

INGREDIENTS:

Bontrae frozen Chicken-Flavor Soy Protein chunks	5 lb.
Cooked chicken, cubed	2 lb.
Celery, sliced	1 lb. 8 oz.
Chinese pea pods or snow peas	1 lb. 8 oz.
Green pepper, julienne	1 lb.
Onions, sliced	1 lb.
Rendered chicken fat or Land O'Lakes Butter	1/2 cup
B & B mushrooms, undrained	2 (6-oz.) cans
Chopped pimientoes	2 oz.
Water	7 cups
Chicken soup base	1/4 cup
Soy sauce	1/2 cup
Cornstarch	3/4 cup

METHOD: Combine all ingredients except water, chicken base, soy sauce and cornstarch in steamjacketed kettle. In saucepot, boil water (reserving 1 cup) with chicken base and soy sauce. Make a smooth paste of reserved 1 cup water and cornstarch. Stir slowly into boiling water. Cook and stir until sauce is thickened and clear. Pour over chicken-vegetable mixture. Stir to mix well. Simmer in steamjacketed kettle for 20 minutes, stirring occasionally. Or, combine all ingredients except sauce mixture in 12- by 20- by 6-in. steamtable pan. Make sauce as directed; stir into vegetable mixture and mix well. Bake at 350F for 40 minutes. Serve with chow mein noodles or rice.

Angostura - Wuppermann Corp.

Make small meat balls as directed for Meat Ball Hoagie, (page 60). Quickly brown meat balls in hot fat, or deep fry for about 2 minutes at 375F. Use as garnish for clear soups such as consomme or bouillon.

CONVENIENCE CHOW MEIN

METHOD: Thaw frozen vegetable chow mein to the "slush" stage, or use drained canned chow mein vegetables. For each pound of vegetable chow mein, stir in 1 (13-oz.) can, drained, Worthington Chicken Soya style.

Heat as directed on package. Serve over rice. Top with dry bacon-flavor soy protein bits.

HASH BROWN CASSEROLE

YIELD: 24 servings

INGREDIENTS:

Dehydrated hash brown potatoes, reconstituted	2 qt.
Paprika	1 Tbsp.
Melted butter	1 cup
Ground beef	2 lb.
Bontrae frozen Beef-Flavor Soy Protein crumbles	1 lb. 12 oz.
Au jus or beef stock	3 cups
Kitchen Bouquet	2 Tbsp
Salt	1 tsp.

METHOD: Brown ground beef in 12- by 20- by 2-in. steamtable pan in oven. Stir occasionally. Combine browned beef with Bontrae; stir to mix. Mix together au jus, Kitchen Bouquet and salt; pour over beef. Mix well. In large bowl toss reconstituted hash browns with paprika and melted butter. Set aside. Spread hash browns evenly over top of beef mixture. Bake at 375F for 20 to 30 minutes.

More Soy Entrees

SUPER PIZZA
YIELD: 4 (18- by 26-in.) pans
INGREDIENTS:

Pillsbury Pizza crust mix	5 lb.
Water	6-1/4 cups
Worthington dry Hamburger-Flavor Soy Granules	1 lb. 6 oz.
Warm water	3 pt.
Pizza sauce	3 to 4 qt.
Ground beeg	3 lb.
Mozzarella cheese, shredded	1 lb.
Parmesan cheese, grated	8 oz.

METHOD: On low speed of mixer, mix pizza crust and 6-1/4 cups water for 1/2 minute. Grease four 18- by 26- in. bun pans with hard or solid shortening. Weigh out 2 lb. pizza crust dough for each pan. Oil hands well. Pat out dough in each pan to form crust.

Rehydrate dry soy protein in 3 pt. warm water for 15 minutes. Spread 3 to 4 cups pizza sauce over each crust. Mix ground beef with rehydrated soy and remaining water. Crumble beef mixture over sauce, using 1-1/2 lbs. for each crust. Then sprinkle each pizza with 4 oz. Mozzarella and 2 oz. Parmesan cheese.

Bake at 450F in conventional oven for 20 minutes; OR

Bake at 350F in a convection oven for about 12 minutes.

VARIATIONS:

Mushroom—Add 8 oz. sliced mushrooms, drained, for each pizza. Put on sauce before crumbling beef mixture over top.

Pepper and Onion—Add 1 cup chopped green pepper for each pizza. Sprinkle over sauce before adding beef mixture.

PIZZA
YIELD: 2 (18- by 26-in.) pizzas
INGREDIENTS:

Pillsbury Pizza crust mix	2 lb. 8 oz.
Water	3 cups plus 2 Tbsp.
Pizza sauce	6 to 8 cups
Ground beef*	2 lb.
Bontrae frozen Beef-Flavor Soy Protein crumbles*	1 lb. 8 oz.
Land O' Lakes mozzarella cheese, shredded	8 oz.
Parmesan cheese, grated	4 oz.

METHOD: Prepare pizza crust mix with water as directed on package. Pat half of dough into each of 2 greased 18- by 26-in. pans. Spread half of sauce over each pizza crust. Combine ground beef and Bontrae; mix thoroughly. Crumble over sauce on pizza crust. Sprinkle with mozzarella and Parmesan cheese.

Bake at 450F in a conventional oven about 20 minutes; OR

Bake at 350F in a convection oven about 12 minutes.

*AMOUNTS FOR SCHOOL LUNCH:
(for 1-1/2 oz. protein per serving, 20 cuts per pan)

Use 1 lb. frozen Bontrae and 3 lb. 8 oz. ground beef; OR
(for 1 oz. protein per serving, 24 cuts per pan)

(Use 14 oz. frozen Bontrae and 3 lb. ground beef.

PIZZA HO HO
YIELD: 1 (18- by 26-in.) pizza
INGREDIENTS:

Pillsbury Pizza crust mix	1 lb. 4 oz.
Water	1-1/2 cups plus 1 Tbsp.
Heinz Barbecue sauce	1-1/2 cups
Water	1-1/2 cups
Salt	1 tsp.
Pimiento-stuffed green olives, chopped	1-1/2 cups
Sharp cheddar cheese, shredded	10 oz.
Ground beef	12 oz.
Bontrae frozen Beef-Flavor Soy Protein crumbles, thawed	8 oz.
Thinly sliced bananas	8 oz.

METHOD: Prepare pizza crust mix with water as directed on package. Pat dough into greased 18- by 26-in. pan. Combine barbecue sauce, water and salt; set aside. On unbaked crust sprinkle olives, cheese, Bontrae and ground beef. Place bananas on surface. Pour barbecue sauce over top.

Bake at 450F in a conventional oven for about 20 minutes;

OR

Bake at 350F in a convection oven for about 12 minutes.

PLAIN LASAGNA

YIELD: 24 servings

INGREDIENTS:

Wide lasagna noodles, uncooked	2 lb.
Boiling water	2 gal.
Salt	2 Tbsp.
Worthington dry Hamburger-Flavor Soy Protein granules	6 oz.
Bulk pork sausage	2 lbs.
Ground beef	1 lb.
Warm water	1-1/2 cups
Onions, chopped	12 oz.
Garlic powder	1/4 tsp.
Salt	2 tsp.
Canned pizza or spaghetti sauce	2 qt.
Creamed small-curd cottage cheese	2 lb.
Mozzarella or "green" cheddar cheese, shredded	1 lb.
Eggs, beaten	3/4 cup
Parmesan cheese, grated	4 oz.

METHOD: Cook noodles in boiling water with 2 Tbsp. salt. Drain and rinse.

In large bowl, mix soy protein granules and 1-1/2 cups warm water. Set aside. Brown sausage and ground beef; remove from pan and add to soy protein. Add onion to drippings in pan; saute until tender. Add onion to sausage mixture. Stir in garlic powder and 2 tsp. salt.

Grease bottom and sides of 12- by 20- by 4-in. steamtable pan. Put in single layer of noodles; spread with 3 cups pizza sauce. Top with second single layer of noodles.

Add eggs to cottage cheese; mix well. Spoon cottage cheese mixture over noodles in pan. Put third single layer of noodles over cottage cheese mixture.

Combine 2 cups pizza sauce with sausage mixture and stir to mix. Spread sausage-sauce mixture over noodles. Top with fourth single layer of noodles.

Mix Parmesan cheese into remaining pizza sauce. Ladle sauce over top of noodles.

Bake at 375F for 40-45 minutes, until sauce bubbles and has a few brown spots on top. Let stand 10 minutes before cutting into portions.

NOTE: To layer noodles, place 1 noodle at corner of pan and lay it flat, lengthwise. Place piece of another noodle at end of first so entire length of pan is covered. Place more noodles alongside first noodles until entire surface of pan is covered.

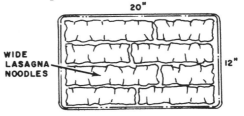

PORCUPINE BALLS

YIELD: 25 servings, 50 balls

INGREDIENTS:

Regular raw rice	13 oz.
Water	5 cups
Ground beef*	2 lb.
Bontrae frozen Beef-Flavor Soy Protein crumbles, thawed*	1 lb. 12 oz.
Dehydrated onion flakes	1/2 cup
Dehydrated diced green peppers	1/4 cup
Durkee Flavor salt	2 Tbsp.
Tabasco	1/2 tsp.
Tomato puree	1 pt.

METHOD: Cook rice in unsalted water. Drain, if necessary. Add remaining ingredients to rice. Form into balls, 2 oz. each, with No. 20 scoop.

Place balls on ungreased baking sheet about 1/2 in. apart. Bake at 400F in a conventional oven for 15 minutes; OR

Bake at 325F in a convection oven for 15 minutes. Serve with catsup.

*AMOUNTS FOR SCHOOL LUNCH: Use 1 lb. frozen Bontrae and 3 lb. 6 oz. ground beef.

FOR GREATER FLAIR: Dolmas—(See picture, page 49)

Use Porcupine Balls, uncooked, as filling for grape leaves. Bake, covered, at 400F for about 20 minutes. Serve with lemon-butter sauce.

Stuffed Cabbage—Steam fresh cabbage outer leaves just to soften or use canned cabbage leaves. Fill with uncooked Porcupine Balls. Roll cabbage leaf around mixture. Place filled leaves in greased 12- by 20- by 2-in. steamtable pan. Cover with 1 qt. Popadic's Spicy Tomato Sauce (page 137). Bake at 400F for 20-30 minutes.

More Soy Entrees

Better Homes and Garden

Stuffed Cabbage Rolls--make mixture for "Porcu-pine Balls" (recipe on page 47) and shape into ovals, 2 oz. each. Wrap in cooked cabbage leaves as shown above. Place seam side down, in greased roasting pan or 12- by 20- by 4-in. steamtable pan. Pour 3 cups of Popadic's Spicy Tomato Sauce over top. Bake at 375F for 30 to 40 minutes.

CHINESE LOBSTER WITH BOWS

YIELD: 50 portions, 1-1/3 cups each

INGREDIENTS:

Dry Pork-Flavor Soy Protein granules	2 lb.
Chicken broth	1-1/2 gal.
Cooked lobster chunks	4 lb.
Ground pork	3 lb.
Vegetable oil	1 cup
Soy sauce	1/2 cup
Brown sugar	1/2 cup
Monsodium glutamate	1/4 cup
Salt	1 Tbsp.
Powdered garlic	1 tsp.
Cold water	2 cups
Kleer Gel(thickener)	1 cup
Bow Knot macaroni	2 lb.
Sliced green onions	1-1/2 cups

METHOD: Rehydrate soy protein in chicken broth at least 20 minutes. Sauté pork in oil until pork is lightly browned. Stir in soy sauce, sugar, monosodium glutamate, salt, garlic. Add rehydrated soy protein and chicken broth. Bring to a boil. Make a smooth paste of cold water and thickener; gradually add to meat mixture, stirring constantly. Simmer, stirring constantly until thickened.

Cook macaroni in boiling, salted water (2 gallons water plus 2 tablespoons salt)until tender, yet firm, 5 to 6 minutes; drain. Serve sauce over macaroni; garnish with green onions.

Durum-Macaroni, HRI Program

Chinese Lobster with Bows (recipe on facing page) extends high cost lobster chunks in flavorful oriental sauce. Sauce is served here on bed of bow knot macaroni in colorful lobster shaped casserole.

More Soy Entrees

Durum-Macaroni, HRI Program

Gourmet Spaghetti Sauce (recipe at right) makes a hearty meal with the traditional accompaniments of al dente spaghetti, wine and "good honest bread." Specialty coffees are the perfect ending. (See recipes on p. 65)

GOURMET SPAGHETTI SAUCE

YIELD: 5 qt., 40 (4-oz.) servings or 25 (6-oz.) servings

INGREDIENTS:

Ground beef	2 lb.
Bontrae frozen Beef-Flavor Soy Protein crumbles, thawed	2 lb.
Sexton's 101 Sauce or All-Purpose Tomato Sauce*	2 qt. plus 1 cup
Water	1 qt.
B in B mushrooms, undrained	2 (6-oz.) cans
Durkee Italian seasoning	5 tsp.
Salt	2 tsp.

METHOD: Brown ground beef; drain off fat. Combine with remaining ingredients in steam jacketed kettle. Cook 15-20 minutes until heated through.

*See recipe on page 134

SPAGHETTI SAUCE

YIELD: 1 gal.

INGREDIENTS:

Ground beef*	2 lb.
Water	2 qt.
Sexton's 101 Sauce or All-Purpose Tomato Sauce**	1 qt.
Tomato puree	1 (29 oz.) can
Durkee Italian seasoning	1/4 cup
Bontrae frozen Beef-Flavor Soy Protein,* crumbles, thawed*	1 lb. 12 oz.

METHOD: Brown ground beef in steamjacketed kettle. Add remaining ingredients except Bontrae. Simmer 45 minutes. Add Bontrae. Simmer 15 minutes longer.

**See recipe on page 134

COST BONUS: Substitute 14 oz. dry hamburger-flavor soy protein granules for the frozen Bontrae.
   Redydrate dry soy protein in 1 qt. warm water.

*AMOUNTS FOR SCHOOL LUNCH: Use 14 oz. frozen Bontrae (or 6 oz. dry soy protein rehydrated in 1-1/2 cups warm water) and 3 lb. ground beef. Has 2 oz. protein per serving, 4 oz. sauce per serving.

POULTRY TETRAZZINI
YIELD: 25 servings
INGREDIENTS:

Regular raw rice	1 lb. 10 oz.
Boiling water	6-1/4 cups
Condensed cream of chicken soup, undiluted	1 (50-oz.) can
Milk	3 cups
Dried parsley flakes	2 Tbsp.
Salt	1 tsp.
Rosemary leaves	1/2 tsp.
Cooked chicken or turkey, cubed	2 lb.
Bontrae frozen Chicken-Flavor Soy Protein chunks, thawed	1 lb. 8 oz.
Parmesan cheese, grated	3 oz.
Paprika	1 Tbsp.

METHOD: Cook rice in boiling water. Drain if necessary. Blend soup, milk, parsley, rosemary and salt. Combine chicken, Bontrae and rice in 12- by 20- by 4-in. steamtable pan. Pour soup mixture over top. Stir to mix well. Mix together cheese and paprika; sprinkle over top of rice mixture. Bake at 400F in a conventional oven for 40 minutes;

OR

Bake at 325F in a convection oven for 35 minutes.

NOTE: 1 lb. spaghetti cooked in 1 gal. boiling water, drained and rinsed, may be substituted for the cooked rice.

SAM'S FAVORITE CURRY
YIELD: About 1 qt. sauce
INGREDIENTS:

Rendered chicken fat	4 oz.
Sliced onions	1 cup
Curry powder	1 Tbsp.
Ground ginger	1 tsp.
Garlic powder	1/2 tsp.
Dry chicken bouillon or base	2 cubes OR 1 Tbsp.
Water	1 qt.
Powdered dry milk	1 cup
Water	1/2 cup
Breading	as required
Cornstarch	2 Tbsp.
Worthington Soyameat, fried chicken style	2 lb. 13 oz. can

METHOD: Melt fat in large saucepan. Saute onions until golden. Add spices and bouillon. Stir in 1 qt. water. Add dry milk. Simmer 15 minutes. Add 1/2 cup water to cornstarch, stirring to make a smooth paste. Stir into sauce. Bring to boil, stirring constantly. Simmer and stir 5 minutes. Then simmer 20 minutes longer.

Drain and rinse soya chicken. Bread, then deep fry at 375F until golden. Serve sauce over soya chicken pieces.

COST BONUS: Substitute 2 lb. frozen chicken-flavor soy protein chunks for the soya chicken. Do not bread or fry; add chunks to sauce along with cornstarch and simmer.

RISI BISI
YIELD: About 1-1/2 gal. (25 servings)
INGREDIENTS:

Regular raw rice	1 lb.
Boiling water	4-1/3 cups
Cooked ham, cubed	2 lb.
Bontrae frozen Ham-Flavor Soy Protein chunks, thawed	1 lb. 4 oz.
Peas	2 qt.
Pimientoes, chopped	4 oz.
Land O'Lakes Butter	4 oz.
Sugar	2 Tbsp.
Dried parsley flakes	1 Tbsp.
Ground cloves	1/4 tsp.
Pineapple juice	5 to 8 oz.
Vermouth or other white wine or water	1 cup
Cornstarch	3 Tbsp.
Water	1 cup

METHOD: Cook rice in boiling water. Combine ham, Bontrae, rice, peas, and pimientoes in a 12- by 20- by 4-in. steamtable pan.

Combine butter, sugar, parsley, cloves and pineapple juice; cook and stir until blended. Make a smooth paste of cornstarch and water. Stir into sauce. Cook, stirring, until thickened and clear. Stir in wine.

Pour over rice mixture. Bake at 400F for 15-20 minutes or until heated through.

More Soy Side Dishes

VEGETABLES ORIENTAL
YIELD: 12 (1-1/4 cup) servings
INGREDIENTS:

Land O'Lakes Butter OR	
Vegetable oil	3/4 cup
Sliced onions	1 lb.
Thinly sliced carrots	1 cup
Thinly sliced celery	1 cup
Green pepper pieces or chunks	1 cup
Pineapple chunks with syrup	2-1/2 cups
Vinegar	1/3 cup
Water	1 pt.
Cornstarch or modified waxy maize starch	1/4 cup
Bontrae frozen Chicken-Flavor or Ham-Flavor Soy Protein chunks, thawed	2 lb.

METHOD: Heat butter in steamjacketed kettle or saucepot. Stir in vegetables and saute until onions are transparent. Add pineapple with syrup and vinegar.

Make a paste of water and cornstarch. Slowly stir into vegetable mixture. Bring to a boil, stirring constantly. Cook and stir until sauce is thick and clear.

Add soy protein. Simmer 15 minutes. If desired, serve with rice, and garnish with chow mein noodles or slivered almonds.

FOR GREATER FLAIR: Drain a 2 lb. 13 oz. can of Worthington soya-meat fried chicken style. Rinse. Bread and deep fry. Cook sauce as directed, but omit soy chunks. Spoon hot sauce over top of fried soya chicken.

SWEET AND SOUR KRAUT
YIELD: 24 (1 cup) servings
INGREDIENTS:

Sauerkraut, drained	1 No. 10 can
Tomato pieces with juice	1 qt.
Brown sugar, firmly packed	6 oz. OR 3/4 cup
Cooked ham, cubed*	2 lbs.
Bontrae frozen Ham-Flavor Soy Protein, chunks, thawed*	1 lb. 12 oz.

METHOD: Combine all ingredients. Mix thoroughly. Turn into a 12- by 20- by 4-in. steamtable pan. Bake at 375F for 35-40 minutes.

*AMOUNTS FOR SCHOOL LUNCH: Use 1 lb. 8 oz. frozen Bontrae and 2 lb. 8 oz. cooked ham. Has 2 oz. protein per serving.

SAUERKRAUT WITH VERVE
YIELD: 24 (3/4 cup) servings
INGREDIENTS:

Sauerkraut, drained	1 No. 10 can
Crushed pineapple with syrup	1 No. 2-1/2 can OR 1 qt.
Cooked ham, cubed	1 lb.
Bontrae frozen Ham-Flavor Soy Protein chunks, thawed	8 oz.

METHOD: Combine all ingredients. Mix thoroughly. Turn into a 12- by 20- by 4-in. steamtable pan. Bake at 400F for about 20-30 minutes, until heated through.

SCALLOPED CORN
YIELD: 18-20 servings
INGREDIENTS:

Dry yam flakes	1 cup
Water	1 cup
Cream-style corn	1 No. 10 can
Condensed cream of celery soup, undiluted	1 pt.
Eggs, beaten	1 cup
Durkee Flavor salt	2 Tbsp.
Prosage Sausage-Flavor Soy Protein, diced or thinly diced	1 lb.
Land O'Lakes Butter	1/3 cup
Coarse cracker crumbs	1 qt.
Bacon-flavored soy bits	1 cup

METHOD: Combine yam flakes and water in 12- by 20- by 4-in. steamtable pan. Stir until smooth. Stir in corn, soup, eggs and salt.

Saute Prosage or fry in deep fat until lightly browned. Drain and add to corn mixture.

Melt butter. Stir in cracker crumbs and bacon-flavored bits. Mix well. Sprinkle crumb mixture over top of corn mixture. Bake at 350F for 45 minutes, until topping is golden.

STROGANOFF

YIELD: 25 (6-oz.) servings

INGREDIENTS:

Ground beef	4 lb.
Bontrae frozen Beef-Flavor Soy Protein crumbles, thawed	2-1/2 lb.
Sliced onions	2 lb.
All purpose flour	1/2 cup
Medium white sauce	1 qt.
Water	1 qt.
Salt	2 Tbsp.
Pepper	1-1/2 tsp.
Sour cream	1 qt.

METHOD: In steamjacketed kettle, brown or saute onions and ground beef. Stir in flour. Combine with remaining ingredients except sour cream in steamjacketed kettle. Cook 15 minutes, stirring occasionally. Stir in sour cream.

At once turn mixture into 12- by 20- by 6-in. steamtable pan.

NOTE: Stroganoff can be held 1 hour on low heat of steamtable.

Stroganoff made of soy protein granules and ground beef is easy on the budget as well as a patron pleaser. Serve with whole vegetables such as green beans or carrots for better appearance on the plate. Or serve in a casserole over wide noodles.

SUPER MAC

YIELD: 25 servings

TOPPING INGREDIENTS:

Dehydrated onion flakes	1/4 cup
Water	1/4 cup
Cornflake crumbs	1-1/2 cup
Bacon flavor soy protein bits	1/2 cup
Salad oil	3 Tbsp.

BASIC INGREDIENTS:

Macaroni, raw weight	1 lb. 9 oz.
Boiling water	1-1/2 gal.
Canned cheese sauce	1 qt.
Water	1 pt.
Powdered dry milk	1 cup
Cooked ham, cubed	1 lb. 4 oz.
Bontrae frozen Ham-Flavor Soy Protein, chunks, thawed	1 lb. 4 oz.

METHOD: Rehydrate onions for Topping in 1/4 cup water for 10 minutes. Then stir in remaining topping ingredients and set aside.

Cook macaroni in boiling water until tender-crisp; rinse and drain well.

Combine cheese sauce with water and dry milk. Cook over low heat, stirring until well blended. Combine with macaroni, ham, and Bontrae in a 12- by 20- by 4-in. steamtable pan. Sprinkle with Topping. Bake at 350F for 45 minutes.

Doug Ross, Photographer

More Soy Side Dishes

Angostura - Wuppermann Corp.

Part of the fun and charm of dining lies in unique tableware. A casserole or crock of chili can be effectively merchandised as a "specialty of the house" in many types of operations-- from coffee shop to cafeteria.

Bacon-Flavor Soy Protein

S&W Fine Foods

Green beans marinated in 2 parts oil and 1 part bottled lemon juice are an even better chilled salad when tossed with bacon-flavor soy protein pieces just before serving.

Things to Do With Bacon-Flavor Soy Protein

The sizes of bacon-flavor soy protein range from crumbles and granules to small, medium or large chips and pieces, plus strips about the size of cooked bacon. Try the bacon-flavor soy protein pieces in a variety of ways.

1 Sprinkle crumbles or chips on top of: scrambled eggs, creamed eggs, curry dishes, salads such as orange slices and onion rings and fresh spinach or mayonnaise potato salad, baked beans, **apple** sauce, green beans, creamy fish chowder, boiled or steamed small potatoes with lemon butter, sauteed mushrooms, creamed onions.

2 Combine bacon-flavor soy protein crumbles with coarse cracker crumbs or corn flake crumbs, using 1 quart soy crumbles to 2 parts cracker crumbs. Add 1/3 cup melted butter and 1/2 teaspoon thyme or rosemary leaves to each 3 cups of crumble/crumb mixture. Sprinkle over top of creamed vegetables; use about 3 cups per 12- by 20-inch steamtable pan. Bake at 400F for 5-10 minutes to crisp the topping.

3 Salads made of:

Avocado slices with green goddess dressing, garnished with bacon-flavor soy protein.

Avocado half, filled with creamy shrimp or lobster salad and a sprinkle of bacon-flavor soy protein.

Tomato slices with vinegar and oil, sprinkled with a mixture of chopped onion and bacon-flavor soy protein.

4 Topping mixture for tossed salad made of equal amounts of grated bread crumbs or croutons, grated cheddar cheese and bacon-flavor soy protein bits.

Tossed salad of lettuce, chopped parsley, radish slices, 1000 Island Dressing and a sprinkle of bacon-flavor soy protein.

5 Sandwiches:

Sprinkle a few bacon-flavor soy protein pieces in egg salad sandwich filling.

Sprinkle bacon-flavor soy protein on the cheese for a grilled cheese special.

Mix bacon-flavor soy protein with cream cheese. Spread on raisin bread for finger sandwiches.

Mix peanut butter, diced pitted dates and bacon-flavor soy protein pieces together; use as sandwich filling on whole wheat bread. Serve with fruit plates.

6 Larger chips of bacon-flavor soy protein are good nibbles to serve as bar snacks.

7 Fry large strips of bacon-flavor soy protein almost crisp, then wrap around whole apricots or purple plums. Use a wooden pick or cocktail skewer to hold. Serve **as a plate** garnish.

Just as in using regular bacon, keep texture and flavor contrasts exciting and delicious to eat. If a crisp texture is expected, sprinkle bacon-flavor soy protein on top and serve at once. Pan fry or deep fry the large strips of bacon-flavor soy protein until of desired crispness. For a chewy texture, mix the pieces of bacon-flavor soy protein with dressing or sauce so they absorb some moisture.

Soy Sandwiches

FANCY DRESS SANDWICH: spread butter over 2 slices white or whole wheat bread. Arrange 2 oz. thinly sliced chicken or ham-flavor soy protein loaf on bread. Warm 1/3 cup red currant jelly or apple-mint jelly in small pan, stirring until jelly is liquid. Spoon jelly over sliced soy protein loaf to glaze. Garnish with toasted almonds.

BUILDER'S BUN: on a toasted sesame bun, spoon hot cheese sauce. Top with fried Stripples bacon-flavor soy protein strips. Add lettuce and a to-mato slice. Arrange thin slices of chicken roll on top. Sprinkle with sauteed chopped onion and sweet relish. Put bun top on. Serve with a peach half.

CASSEROLE SANDWICH: serve in small oval casseroles or au gratin dishes. Use 1 cup cream sauce or canned condensed cream of mushroom soup, heated, with 1 cup toasted bread cubes and 3/4 cup sauteed chunks of chicken- or ham-flavor soy protein. Garnish with a French fried onion ring.

Advisory Council for Jams, Jellies and Preserves

A sizzling sandwich platter takes on new dimensions with hot glazes made from colorful jellies. Here, red currant or apple mint jelly is the tart topping added to a ham-flavor soy protein loaf sandwich. Also good when chicken-flavor soy protein loaf is the filling.

More Soy Sandwiches

U. S. Brewers Association

Monte Cristo Burger—Hamburgers made of Meat Loaf II (recipe on page 35) are pre-cooked. Place each burger between 2 slices of Vienna bread along with slice of onion. Make batter of equal parts beer and egg; beat until frothy then dip whole sandwich in quickly. Grill or deep fry at 375F until golden. Can also be called Batter Burger.

BARBECUE BEEF

YIELD: About 5 qts., 50 servings

INGREDIENTS:

Ground beef*	2 lb. 8 oz.
Bontrae frozen Beef-Flavor Soy Protein crumbles, thawed*	2 lb.
Dehydrated onion flakes	1 cup
Durkee Barbecue spice	1 cup
Sexton's 101 Sauce or All-Purpose Tomato Sauce**	3 qts.
Water	1 pt.

METHOD: Brown ground beef in steamjacketed kettle. Add remaining ingredients, stir until well mixed. Cook 15 minutes, stirring occasionally.

**See recipe on pages 134-135

*AMOUNTS FOR SCHOOL LUNCH: Use 1 lb. frozen Bontrae and 3 lb. 2 oz. ground beef, OR substitute dry soy protein for the frozen crumbles; use 6 oz. dry hamburger-flavor soy protein granules, rehydrate in 1 pt. warm water for 15 minutes. Has 1 oz. protein per serving.

SLOPPY JOE

YIELD: About 1-1/2 gal.

INGREDIENTS:

Ground beef*	3 lb.
Chopped onion	2-1/2 cups
Worthington dry Hamburger-Flavor Soy Protein granules	12 oz.
Water	3 cups
Brown sugar	8 oz.
Dried parsley flakes	2 Tbsp.
Salt	4 tsp.
Pepper	1 tsp.
Tomatoes	1 (29-oz.)can
Tomato puree	1 (29-oz.)can

METHOD: Brown beef and onions. Meanwhile, in separate pan or bowl, rehydrate soy granules in water.

Add sugar and seasonings to browned beef. Add tomatoes and tomato puree. Simmer about 45 minutes. Stir in soy granules along with any excess water. Simmer 15 minutes longer.

*AMOUNTS FOR SCHOOL LUNCH: Each 4 oz. serving contains 1-1/4 oz. protein.

Pizza is a favorite of big and little kids. Try baking it in half-size bun pans, each with a different topping variation. Shown here is the ground beef-soy protein mixture formed into small balls.

CHICKEN SANDWICH FILLING

YIELD: 6 lb. 8 oz.

INGREDIENTS:

Cooked chicken, ground	2 lb. 4 oz.
Bontrae frozen Chicken-Flavor Soy Protein chunks, thawed and ground	1 lb.
Canned tartar sauce	3 pt.

METHOD: Combine all ingredients; stir by hand, or use paddle on low speed of electric mixer, until well blended. Use a No. 16 scoop for each sandwich.

THIN SLICED CUCUMBERS

YIELD: 6 servings

INGREDIENTS:

Tarragon vinegar	3/4 cup
Water	1/4 cup
Sugar	2 Tbsp.
Salt	1 tsp.
Cucumber, thinly sliced	2-1/2 cups
Sesame seeds, toasted	1/2 cup

METHOD: Combine vinegar, water, sugar and salt in shallow bowl. Add cucumber slices. Toss lightly to mix. Refrigerate for at least 4 hours before serving. Garnish with sprinkle of sesame seeds.

FOR GREATER FLAIR: Serve in small souffle cups as a garnish or side salad with sandwiches in a basket.

Also makes an attractive sandwich garnish if cucumber is sliced lengthwise, marinated, then rolled and skewered on top of sandwich.

More Soy Sandwiches

Try Meat Ball Hoagie merchandising ideas: serve open-face, substitute heated pizza sauce for chili sauce and sprinkle with Parmesan; feature as the "Italian Meal on a Bun." Offer 2 hoagies as a lovers' lunch along with a soft drink or wine. Or substitute heated pork and beans for chili sauce and call it a "Big John."

MEAT BALL HOAGIE
YIELD: 35 sandwiches
INGREDIENTS:

Meat Loaf II	1 recipe p.42
Sauerkraut, well drained	2 lb.
Vega-Links All-Soy Hot Dogs, cut in 1-in. pieces	1/2 of 4 lb. 12 oz. can
Chili sauce	as needed
Hoagie buns	35

METHOD: Shape Meat Loaf II mixture into 3/4 oz. balls. Place in greased roasting pan. Bake at 400F about 15 minutes, turning once or twice.

For each sandwich, use:
- 1 hoagie bun, lined with 1 oz. sauerkraut
- 3 meat balls
- 2 or 3 pieces Vega-Links

BLT in a Basket—features slices of American cheese and tomato with lettuce and Stripples bacon-flavor soy protein strips. Serve grilled or toasted.

Wheat Flour Institutue
National Sandwich Idea Headquarters

FRANK BURGERS

YIELD: about 65 burgers
INGREDIENTS:

Vega-Links All-Soy Hot Dogs, drained	1 (4 lb. 12 oz.) can
Meat Loaf II	1 recipe (p. 42
Barb's Mayonnaise Stuff	1 recipe (below)

METHOD: Thinly slice hot dogs. By hand, mix into Meat Loaf II mixture. Form into rounds or patties, using No. 12 scoop. (Burgers should weigh 2-1/2 - 3 oz. each.) Grill or roast as for regular hamburgers. Serve on buns with Barb's Mayonnaise Stuff.

BARB'S MAYONNAISE STUFF

YIELD: about 2 qt.
INGREDIENTS:

Mayonnaise	1 qt.
Catsup	1 pt.
Sweet relish	1 cup
Prepared mustard	1/2 cup

METHOD: Combine all ingredients; mix well. Store in refrigerator.

SANDWICH PATTIES

YIELD: about 18 lb.
INGREDIENTS:

Unflavored soy grits	1 lb. 8 oz.
Warm water	1 qt.
Lean beef (chuck)	10 lb.
Beef suet	2 lb. 8 oz.
Salt	1/2 cup
Ice with water	1 qt.

METHOD: Rehydrate dry soy grits in warm water for 20 minutes. In meat grinder, combine all ingredients. Grind about 30 seconds, until mixture forms a compact mass. Form into patties.

VARIATIONS: Hamburgers form the basis for a wild variety of "ethnic" sandwiches.

Pizza burger--add a slice of mozzarella and spoon on spaghetti or pizza sauce.

Taco burger - Sprinkle on shredded lettuce, chopped tomato and onion; spoon on taco sauce.

Burrito burger - spoon on refried beans and shredded cheddar cheese; top with sour cream.

Vegetable burger - add slices of tomato, onion and pickles; top with lettuce.

Archer-Daniels-Midland

Hamburgers are Americans' No. 1 favorite, according to the Institutions/VFM menu census report. Using soy protein reduces shrinkage and makes patties juicier. Meat Patties made with un-flavored or beef-flavor soy protein (recipe on page 64) offer limitless variety with just enough **change** to keep people coming back for more and more.

Soy Breads

Wheat Flour Institute

Swirl Rolls (recipe on page 64) can be varied by baking in greased muffin pan and adding a sprinkle of grated Parmesan when hot from the oven.

CORN MUFFINS OLÉ
YIELD: 80 (1-1/2 oz.) muffins
INGREDIENTS:

Corn muffin mix	5 lb.
Dehydrated onion flakes	2 cups
Chili powder	2 Tbsp.

Bontrae frozen Beef-
  Flavor Soy Protein
  chunks, thawed
	12 oz.

Water	5-1/2 cups

METHOD: Combine all ingredients in mixer bowl. Mix with paddle as directed on muffin mix package. Spoon into greased muffin pans.

Bake at 400F in a conventional oven for 15-18 minutes; OR

Bake at 325F in a convection oven for 7-10 minutes.

TAVERN CORN MUFFINS
YIELD: About 80 (1-1/2 oz.) muffins
INGREDIENTS:

Corn muffin mix	5 lb.
Process American cheese, shredded	4 oz.

Bontrae frozen Beef-
  Flavor Soy Protein
  crumbles, thawed
	6 oz.

Beer	2-1/2 cups

METHOD: Combine all ingredients in mixer bowl. Mix with paddle as directed on muffin mix package. Spoon into greased muffin pans.

Bake at 400F in a conventional oven for 15-18 minutes; OR

Bake at 325F in a convection oven for 7-10 minutes.

CRUSTY BUNS
YIELD: 6 dozen
INGREDIENTS:

Pillsbury Hot roll mix	5 lb.
Soy flour or powdered soy protein	1 cup
Yam flakes	1 cup
Condensed tomato soup, undiluted	1 (50 oz.) can
Water	1 pt.
Bacon-Flavor soy pro- tein bits or granules	4 oz.
Land O' Lakes butter, melted	as needed

Parmesan cheese,
  grated
  OR Sesame seeds
	as needed

METHOD: Combine all ingredients in mixer bowl; mix with dough hook for about 10 minutes. In warm place, let rise in greased bowl until dough is doubled in bulk. Punch down. Roll into balls, about 1-3/4 oz. each; OR

Portion dough into greased muffin pans. Let rise until doubled. Brush tops with melted butter. Sprinkle with Parmesan or sesame seeds.

Bake at 400F in a conventional oven for 15-20 minutes; OR Bake at 325F in a convection oven 12-15 min.

SOOFELS
Soofels are almost like souffle—airy, flavorful, good with sauces. Soofels are also like spoon bread—delicate, fun to eat, delicious with melted butter.

YIELD: 12 (1-cup) servings
INGREDIENTS:

Scalded milk	3 cups
Quaker Instant Grits	8 (0.8 oz.) packets
Land O' Lakes Butter	1/3 cup
Egg yolks, beaten	2/3 cup
Egg whites	1-1/3 cup
Heinz malt vinegar	1 Tbsp.

Bontrae frozen Beef-
  Flavor Soy Protein
  crumbles, thawed
	5 oz.

Bacon-Flavor soy bits	1 oz.
Parmesan cheese, grated	as needed

METHOD: Butter sides and bottom of 12 individual (2-in. deep, 5-in. diameter) casseroles; dust with grated Parmesan cheese. Stir hot milk into grits. Add butter and stir until melted. Stir small amount of grits into egg yolks; add egg yolk mixture to grits and stir until well blended. Beat egg whites until frothy; add vinegar and beat until stiff peaks form. Fold egg whites, Bontrae and bacon bits into grit mixture. Pour into individual casseroles. Bake at 375F in a conventional oven for 25-30 minutes; OR Bake at 300F in a convection oven for 20 minutes. Serve hot.

SERVE Soofels with a sauce or syrup: Chinese Sweet-Sour Sauce, orange marmalade thinned with lime juice, chutney, cheese sauce thinned with beer, au jus with a little currant jelly, Hollandaise, a mushroom brown sauce or Bordelaise, sour cream with chives or horseradish.

More Soy Breads

Wheat Flour Institute

Swirl Rolls—pan rolls with hearty flavor of cream cheese, onions and bacon are taste treat with butter and even better with a bit of honey too. A quick brush with egg wash before baking holds on sesame seeds which are sprinkled on top.

SWIRL ROLLS
YIELD: about 90
INGREDIENTS:

Pillsbury Hot roll mix	5 lb.
Water	5 cups
Cream cheese	12 oz.
Eggs	2
Dehydrated onion flakes	2/3 cup
Bacon-Flavor soy protein bits	6 oz.

METHOD: Prepare hot roll mix with water as directed on package. Let rise. Punch dough down and turn out onto floured board. Roll or pat out to 1-in. thickness.

Blend cream cheese, eggs and onion flakes until well mixed. Spread half of cheese mixture over dough. Sprinkle on half of soy bits. Roll dough loosely, jellyroll fashion. Pat or roll out again to 1-in. thickness. Spread with remainder of cheese mixture. Sprinkle with remaining soy bits. Fold dough over and knead lightly. Shape into 1-1/2 oz. rolls. Place on greased baking pans. Proof for about 1 hour, or until light and about double in size. Bake at 400F for 13-18 minutes, until golden.

*SCHOOL LUNCH AMOUNTS: Each roll contains 1/2 oz. protein.

FRUIT KUCHEN
YIELD: 2 (13- by 18-in.) pans
INGREDIENTS:

Biscuit mix	2 lb. 8 oz.
Syrup (from pears and peaches)	2-1/4 cups
Land O'Lakes Butter, softened	8 oz.
Sliced pears, drained	1 No. 2-1/2 can
Sliced peaches, drained	1 No. 2-1/2 can
Dark sweet pitted cherries, drained	1/2 of No. 10 can
Milk	5-1/2 cups
Eggs	1-1/3 cups
Sugar	1/2 cup
Vanilla	1 Tbsp.
Salt	1 tsp.
Unflavored No. 120 minced ADM Soy Protein	5 oz.

METHOD: Blend biscuit mix, syrup and butter on low speed of mixer, using the paddle. Divide dough into 2 greased 13- by 18-in. pans; spread evenly over bottom. Arrange pears, peaches and cherries over dough. Combine milk, eggs, sugar, vanilla and salt; beat until frothy. Add soy protein and pour over fruit. Bake at 375F in a conventional oven for 20-25 minutes;

OR

Bake at 300F in a convection oven for 15-20 minutes.

NOTE: To use full 5-lb. box of biscuit mix, double the recipe.

SPECIALTY COFFEES

Start with extra strength coffee made by dissolving 1 single-serve packet Maxwell House instant coffee in 1 cup regular coffee.

1 Brandied Coffee: for 2 servings use 1-1/2 cups extra strength coffee, sweeten with 2 Tbsp. sugar. Pour 1 shot brandy into each cup when served.

2 Cafe Brulot: for 2 servings use 2 cups extra strength coffee, 6 whole cloves, 1/2 stick cinnamon, 2 Tbsp. sugar, slivered orange and lemon peel. Let steep 5 minutes. Serve flaming by carefully pouring a teaspoon of brandy or flambe on top of coffee in each cup, then ignite; add 1 shot cognac to each cup after flame is gone.

3 Cappucino: for 2 servings use 1 cup extra strength coffee, mix with 1 cup hot milk. Serve in tall mugs with cinnamon sticks.

4 Coffee Viennoise: for 2 servings use 2 cups extra strength coffee and 1 shot light rum. Serve in tall footed cups with a dollop of whipped topping on each.

5 Demitasse: for 2 servings use 1 cup extra strength coffee. Serve in tiny coffee cups.

6 Irish Coffee: for 2 servings use 2 cups extra strength coffee, 1 Tbsp. sugar and 1 shot Irish whiskey. Stir. Serve in Irish Coffee mugs; top each with dollop of whipped topping.

General Foods Kitchens

Favorite endings include whipped topping—on coffee as well as on desserts. Coffee Wafers (recipe on page 66) are sweet but not too sweet—perfect to serve with specialty brews such as Irish Coffee.

More Soy Desserts

COFFEE WAFERS

YIELD: About 4 doz.

INGREDIENTS:

Land O'Lakes Butter	4 oz.
Granulated sugar	2/3 cup
Brown sugar, firmly packed	1/3 cup
Vanilla	1 tsp.
Egg	1
Baking powder	1 tsp.
Salt	1/2 tsp.
Brewed coffee	1/2 cup
Dry instant coffee	1 individual packet
Soy flour or powdered soy protein	1 cup
All purpose flour	1-1/2 cups

METHOD: Cream butter, sugars, and vanilla. Beat in egg, baking powder and salt. Add liquid and dry coffee; blend until smooth. Beat in soy flour and all purpose flour; mix thoroughly. Drop by teaspoonfuls onto greased baking sheets. Press with bottom of glass or spatula until cookies are 1/4-in. thick.

Bake at 375F for 8-10 minutes, until edges are browned. Cookies become crisp as they cool. Sprinkle with chocolate shot or confectioners' sugar, if desired.

CHOCOLATE CHIP COOKIES

YIELD: About 8 dozen

INGREDIENTS:

Brown sugar	1 lb.
Granulated sugar	1 cup
Land O'Lakes Butter, softened	1 lb.
Eggs	3/4 cup
Vanilla	1 Tbsp.
Soy flour or powdered soy protein	1-1/2 cups
All purpose flour	3 cups
Baking soda	1 Tbsp.
Salt	2 tsp.
Chocolate chips	1-1/2 lb.
Chopped walnuts or pecans	8 oz.

METHOD: On high speed of mixer, cream sugars and butter. Add eggs and vanilla and beat until well blended. At low speed, mix in soy flour, all purpose flour, soda and salt. At low speed, or by hand, mix in chocolate chips and nuts. Drop by tablespoons or a No. 40 scoop (or 1 oz.) on greased baking sheet. Bake at 375F in a conventional oven for 10-15 minutes;

OR

Bake at 275F in a convection oven for 12 minutes.

DATE BARS

YIELD: 2(13- by 18-in.)pans

TOPPING INGREDIENTS:

All purpose flour	1-1/2 cups
Unflavored No. 120 minced ADM Soy Protein	12 oz.
Brown sugar	6 oz.
Granulated sugar	6 oz.
Land O'Lakes Butter	1 cup

FILLING INGREDIENTS:

Bordo Diced dates	3 lb.
Water	1 qt.
Granulated sugar	12 oz.
ReaLemon Lemon juice	1/4 cup
Salt	1/2 tsp.

CRUST INGREDIENTS:

Land O'Lakes Butter, soft	12 oz.
Eggs	1-1/4 cups
Brown sugar	1 lb. 8 oz.
Baking soda	1 Tbsp.
Salt	1-1/4 tsp.
All purpose flour	1-1/2 lb.
Soy flour or powdered soy protein	1 cup

METHOD: TOPPING: Combine flour, soy protein, sugars and butter; mix until crumbly; set aside.

FILLING: In saucepan, combine dates, water, sugar, lemon juice and salt; cook until thick; cool. Set aside.

CRUST: At medium speed on mixer beat butter, eggs, brown sugar, soda and salt until well blended; slowly beat in flours to form soft dough. Spread half of dough in each 13- by 18-in pan.

Spread half of date filling over dough in each pan. Top each pan with half of crumb topping mixture. Bake at 375F in a conventional oven for 35 minutes;

OR

Bake at 300F in a convection oven for about 25 minutes.

RAISIN ROCKS

YIELD: About 7 doz.

INGREDIENTS:

Brown sugar	1 lb.
Land O'Lakes Butter, softened	12 oz.
Eggs	3/4 cup
Milk	2 Tbsp.
Soy flour or powdered soy protein	1-1/2 cups
All purpose flour	2-1/2 cups
Baking soda	1 tsp.
Ground cinnamon	1 tsp.
Salt	1 tsp.
California seedless raisins	1 lb.
Chocolate chips	12 oz.

METHOD: On high speed of mixer, cream brown sugar and butter. Add eggs and milk and beat until well blended. On low speed, blend in soy flour, flour, soda, cinnamon and salt. On low speed, or by hand, mix in raisins and chocolate chips.

Drop by tablespoons or a No. 40 scoop (or 1 oz.) on greased baking sheet. Bake at 375F in a conventional oven for about 15 minutes;

OR

Bake at 300F in a convection oven for about 10 minutes.

PURPLE PLUM CRISP

YIELD: 24 servings

INGREDIENTS:

Purple plums	1 No. 10 can
ReaLemon Lemon juice	1/4 cup
Brown sugar, firmly packed	1/2 cup
Cornstarch	1/3 cup
Water	1/2 cup
Ground cinnamon	1-1/2 tsp.
Salt	1 tsp.
Ground nutmeg	3/4 tsp.
Ground cloves	1/4 tsp.
Vanilla	1 Tbsp.

TOPPING:

Land O' Lakes Butter melted	1-1/2 cups
Rolled oats	1 cup
Wheat germ	1 cup
Dry fine-minced Soy Grits or fine-minced Soy Protein	1 cup
Brown sugar, firmly packed	1 cup
Ground cinnamon	1 Tbsp.
Salt	1 tsp.

METHOD: Drain plums, reserving 3 cups of syrup. Cut plums in half and remove pits. Place plums in 12- by 20- by 2-inch pan.

Bring 3 cups of plum syrup, lemon juice, and brown sugar to a boil. Make a smooth paste of cornstarch and water; slowly stir into boiling liquid. Add cinnamon, nutmeg, salt and cloves. Cook and stir until mixture is thickened. Pour over plums. Stir to mix.

Combine all ingredients for Topping; mix well. Sprinkle over plums. Bake at 375F for 20 minutes. Serve hot with whipped topping.

DATE-FRUIT BREAD

YIELD: 6 loaves, about 1 lb. 10 oz. each.

INGREDIENTS:

Fruit cocktail, undrained	1 No. 2-1/2 can
Water	as needed
Pillsbury Date bread mix	5 lb.
Soy flour OR Powdered soy protein	1 cup

METHOD: Drain fruit cocktail and save the syrup. Add water to the syrup to make 5-1/2 cups liquid. Put liquid in mixer bowl. Add remaining ingredients; scrape down sides of bowl. With paddle, mix on low speed for 1/2 minute longer. Divide into 6 greased and floured 9- by 5-in. loaf pans.

Bake at 350F for about 1 hour. Turn loaves out onto rack to cool. Slice. Serve toasted with butter.

MINCEMEAT PIE FILLING

YIELD: Filling for 2-crust 9-in. pie

INGREDIENTS:

Cider	3/4 cup
ADM No. 120 minced dry Beef-Flavor or un-flavored soy protein	3 oz.
Moist mincemeat	1 lb. 12 oz.

METHOD: Combine cider and soy protein. Let rehydrate for 10 minutes. Stir in mincemeat. Fill unbaked 9-in. pie shell, top with second crust and bake as usual.

Soy Miscellany

Soda Pop—the cola drinks and other sugar-sweetened soft drinks are good liquids for rehydrating beef-flavor soy protein products because the meat flavor is enhanced.

Spices—usually more, than less, is the rule with spices used in soy protein dishes. Blends of spices, such as: apple pie spice, barbecue spice, curry powder, flavor salt, Italian seasoning and the salad spice mixes are all fun to use with a liberal hand. As suggested elsewhere in this book, the "second day" spicy food with soy protein may not hold well. For these dishes, spice a little during preparation and add more before serving—if needed. Soy protein dishes usually freeze well, but some adjustments are often needed in seasoning. Anise, caraway seed, rosemary and sage are best used sparingly, especially in all-soy items.

Wine—generally the sweet wines are preferable to hard liquor or beer in soy product cookery. Short marinating times (15 to 30 minutes) and boiling to evaporate most of the alcohol produce the best recipes. Sweet liqueurs also give a very nice flavor in soy cookery.

TOPPINGS FOR VEGETABLES

Make cornicopias or rolls of sliced Chicken-Flavor or Ham-Flavor soy protein loaf or roll. Use to garnish vegetables served buffet style or pans on the steamtable.

Heat slices of Chicken-Flavor or Ham-Flavor soy protein loaf or roll in steamer. (This product dries out very quickly.)

Saute sliced onions and Bacon-Flavor soy protein; top mashed potatoes.

PEANUT BUTTER COOKIES
YIELD: About 8 doz.
INGREDIENTS:

Ingredient	Amount
Brown sugar	1 lb.
Granulated sugar	1 lb.
Shortening	1 lb.
Peanut butter	1 lb. 4 oz.
Eggs	1 cup
Vanilla	1 tsp.
Soy flour or powdered soy protein	1 cup
All purpose flour	1 lb.
Baking soda	1 Tbsp.
Baking powder	1 tsp.
Salt	2 tsp.

METHOD: On high speed of mixer cream sugars with shortening and peanut butter until well blended. Beat in eggs and vanilla. Blend in remaining ingredients on low speed of mixer or by hand. Form into 1-in. balls, use No. 40 scoop if desired.

Place on greased baking sheet and flatten with bottom of a glass dipped in sugar or with fork. Bake at 375F in a conventional oven for about 10 minutes;

OR

Bake at 300F in a convection oven for about 8 minutes.

*AMOUNTS FOR SCHOOL LUNCH: Form dough into 1-1/2 in. balls. Bake flattened cookies at 375F for about 12 minutes in in conventional oven. Yield is about 5-1/2 doz. Has 1/2 oz. protein per cookie, made with soy flour.

Suppliers List

The following companies process soy products. Most sell several types, sizes and flavors of soy products. Some make only ingredients (for sale to food manufacturers).

Soy Specialties Division
Archer Daniels Midland Co.
4666 Faries Parkway
Decatur, Ill. 62525

Miles Laboratories
1127 Myrtle
Elkhart, Ind. 46514

Cargill, Inc.
Cargill Building
Minneapolis, Minn. 55402

Ralston Purina Co.
Checkerboard Square
St. Louis, Mo. 63199

Chemurgy Division
Central Soya
1825 N. Laramie
Chicago, Ill. 60639

A.E. Staley Manufacturing Co.
P. O. Box 151
Decatur, Ill. 62525

Far-Mar-Co., Inc.
960 N. Halstead
Hutchinson, Kansas 67501

Swift Chemical Co.
1211 W. 22nd St.
Oak Brook, Ill. 60521

Food Service & Protein Products Div.
General Mills, Inc.
9200 Wayzata Blvd.
Minneapolis, Minn. 55440

H. B. Taylor Co.
4830 S. Christiana Ave.
Chicago, Ill. 60632

Worthington Foods, Inc.
900 Proprietors Rd.
Worthington, Ohio 43085

Griffith Laboratories
1415 W. 37th St.
Chicago, Ill. 60609

Worthington Foods, Inc.

INDEX